Master Math:
Business and Personal Finance Math

Mary Hansen

Course Technology PTR

A part of Cengage Learning

COURSE TECHNOLOGY
CENGAGE Learning™

Australia • Brazil • Japan • Korea • Mexico • Singapore • Spain • United Kingdom • United States

COURSE TECHNOLOGY
CENGAGE Learning™

Master Math: Business and Personal Finance Math

Mary Hansen

Publisher and General Manager, Course Technology PTR:
Stacy L. Hiquet

Associate Director of Marketing:
Sarah Panella

Manager of Editorial Services:
Heather Talbot

Marketing Manager:
Jordan Castellani

Senior Acquisitions Editor:
Emi Smith

Project Editor and Copy Editor:
Kim Benbow

Technical Reviewer:
David Lawrence

Interior Layout Tech:
Judy Littlefield

Cover Designer: Mike Tanamachi

Indexer: Larry Sweazy

Proofreader: Sue Boshers

Library of Congress Control Number: 2011922401

ISBN-13: 978-1-4354-5788-1

ISBN-10: 1-4354-5788-9

Course Technology, a part of Cengage Learning
20 Channel Center Street
Boston, MA 02210
USA

Cengage Learning is a leading provider of customized learning solutions with office locations around the globe, including Singapore, the United Kingdom, Australia, Mexico, Brazil, and Japan. Locate your local office at: **international.cengage.com/region.**

Cengage Learning products are represented in Canada by Nelson Education, Ltd. For your lifelong learning solutions, visit **courseptr.com.**

Visit our corporate website at **cengage.com.**

Printed by RR Donnelley. Crawfordsville, IN. 1st Ptg. 04/2011

Printed in the United States of America
1 2 3 4 5 6 7 13 12 11

To my children—

It is my prayer that you will always embrace learning and cultivate a love of mathematics.

Notice to the Reader:

Acknowledgments

It is with great thankfulness that I reflect on the many people who have supported and encouraged me both in mathematics and in this project. Thank you, Mr. Mayer, for opening my eyes in junior high school to the usefulness of mathematics in solving real problems in the real world. I am grateful for my Dad and Mom for encouraging me, especially to you, Dad, for sharing with me your talent for math and reasoning. I am grateful to my high school geometry teacher, Mrs. Benson, and my college advisor and mathematics professor, William Trench, who taught many of my mathematics classes and also allowed me to work on his life's contribution to mathematics. Both challenged me to stretch myself and learn more.

I never outgrew a childhood dream to be a teacher, and I am so thankful that in my teaching jobs I was encouraged by many to explore new ways to teach mathematics. Specifically, I appreciate my college education professor, David Molina; my mentor teacher, Laurie Bergner; and principals, Chula Boyle and Joanne Brookshire, who supported my dream that all students can learn mathematics.

I am thankful for Eve Lewis and Enid Nagel from South-Western Publishing who believed in my writing and teaching abilities and chose me, an unpublished writer, to be on the authorship team of the South-Western Algebra, Geometry, and Algebra 2 series, beginning my journey in the world of educational publishing. Thank you Emi Smith and Kim Benbow for your work on this project and your patience as we sorted out various obstacles.

Finally, I would be remiss if I did not thank my wonderful husband, who has supported me and believed in me through all the years and different projects. You have been loving and patient. You have always encouraged me to do my best and take the next step. Thank you, my love.

About the Author

Mary Hansen has taught K-12 and post-secondary mathematics and special education in three states. She has travelled the United States extensively, doing teacher workshops on effective teaching strategies and effective mathematics teaching with two different educational consulting firms. Hansen received a Master of Arts in teaching and a bachelor's degree in mathematics from Trinity University in San Antonio, Texas. She is the author of *Business Mathematics, 17th Edition* and the co-author of *South-Western Algebra 1: An Integrated Approach*, *South-Western Geometry: An Integrated Approach*, and *South-Western Algebra 2: An Integrated Approach* (all from South-Western Publishing).

Table of Contents

Chapter 10: Business Costs

Chapter 11: Sales

Chapter 12: Inventory

Chapter 13: Financial Statements and Ratios

Introduction

Many believe that there are two kinds of people in this world—those who can do math and those who can't. I respectfully disagree. While there are certainly people who have a talent or a gift for mathematics, all people are capable of learning and applying mathematics.

In our society, avoiding mathematics is not only a difficult task, but a risky one. Paychecks, taxes, a checking account, credit cards, and loans are an everyday part of the personal finance and business world. To choose to learn nothing is to be at the mercy of others and our own uninformed decisions.

The most common question posed to a math teacher is, "When am I ever going to have to use this?" You will not have to ask that question when working through this book because its content is required to understand, and make good decisions about, personal finance and business. This is the math you can't afford to miss out on!

It has always been my goal as a teacher to make mathematics relevant and understandable to students. While I wish I could sit beside you as you work through this content, I have done my best to explain the topics in everyday language and to show shortcuts and tips that will help you understand the process and perform the mathematics more quickly and easily.

Master Math: Business and Personal Finance Math is designed as a reference and resource tool. It might be used by a student taking a business math or personal finance course who wants another resource to supplement a textbook. It might be used to refresh skills that have gotten rusty or to learn new skills to assist a person in managing personal finances or working in the business world.

The book begins with a review of basic math skills that will be encountered throughout the book; it then moves into topics that impact both personal finance and business, including

- Calculating paychecks and taxes.
- Maintaining bank accounts, investments, and insurance.
- Managing credit.
- Making major purchases.
- Creating and managing a budget.

The remainder of the book focuses on topics that are specific to a business, such as

- Managing costs, sales, and marketing.
- Creating and analyzing business statements.

Depending on your needs, you may choose to first work through all of the Chapter 1 math review to prepare for the skills utilized in the remaining chapters, or you may find that you only need to brush up on a few skills from the first chapter. Alternatively, you may choose to come back to the first chapter as you work through the rest of the content and encounter skills on which you need a refresher.

The remaining chapters cover content typical to a business math or personal finance class, and encompass skills that will prepare you to organize, understand, and calculate with numbers so that you can make good decisions in both personal and business settings.

Each chapter is broken into named sections so that you can find a specific topic easily. Each section has one or more example problems, along with several practice problems, so that you can test your skills. An appendix provides not only the answers for all practice problems, but also the mathematical steps used to arrive at the answers, so you can check your work each step of the way.

Chapter

1

Math Review

Basic math skills are needed in order to calculate everything from personal checkbook balances to entries on business financial statements. This chapter is a review of the basic math skills that are utilized throughout this book. If needed, you can work through all of Chapter 1 to prepare for the math you will encounter. You may find that you only need to brush up on a few of the skills in this chapter, or you may prefer to come back to this chapter as you work through other chapters and encounter skills on which you need a refresher.

1.1 Fractions

A *fraction* is a way of representing a whole divided into equal parts. In a fraction, the number on the top is called the *numerator*, and the number on the bottom is called the *denominator*.

$$\text{Fraction} = \frac{\text{Numerator}}{\text{Denominator}} = \frac{\text{Number of Equal Parts Chosen}}{\text{Total Number of Parts}}$$

Equivalent Fractions

Any given fraction has an endless number of fractions that are equivalent. Consider dividing a pie into two pieces. One piece is $\frac{1}{2}$ of the pie. But if the pie is divided into six pieces, then $\frac{3}{6}$ is also $\frac{1}{2}$ of the pie. Or the pie can be divided into eight pieces, and $\frac{4}{8}$ is still $\frac{1}{2}$. Each of these fractions is an *equivalent fraction*.

Example: Find a fraction equivalent to $\frac{3}{4}$.

Solution: Multiply or divide the numerator and denominator of the fraction by the same number.

$$\frac{3}{4} = \frac{3 \times 2}{4 \times 2} = \frac{6}{8}$$

$\frac{6}{8}$ is equivalent to $\frac{3}{4}$.

PRACTICE PROBLEMS

1.1   Find a fraction equivalent to $\frac{2}{3}$ by multiplying the numerator and denominator by 4.

1.2   Find a fraction equivalent to $\frac{10}{12}$ by dividing the numerator and denominator by 2.

A fraction is said to be reduced to *lowest terms* if no number other than 1 can be divided evenly into both the numerator and denominator.

Example: Reduce $\frac{25}{30}$ to lowest terms.

Solution: Divide the numerator and denominator by the largest number that will divide evenly into both.

5 will divide into 25 and 30.

$$\frac{25}{30} = \frac{25 \div 5}{30 \div 5} = \frac{5}{6}$$

$$\frac{25}{30} = \frac{5}{6}$$

Practice Problems

1.3 Reduce $\frac{4}{12}$ to lowest terms.

1.4 Write $\frac{18}{42}$ in lowest terms.

Other fractions that can be equivalent are improper fractions and mixed numbers. An *improper fraction* is a fraction where the numerator is larger than the denominator. An improper fraction can be turned into a whole number and fraction, called a *mixed number*.

To change an improper fraction into a mixed number, divide the numerator by the denominator to find the whole number. The fractional part of the mixed number is formed by the remainder from the division as the numerator over the original denominator.

$$\text{Improper Fraction} = \frac{16}{7} = 2\frac{2}{7} = \text{Mixed Number}$$

Example: Write $\frac{14}{3}$ as a mixed number.

Solution: Divide 14 by 3 and find the remainder. Place the remainder on the top of the fraction, with the divisor as the denominator.

$$
4\frac{2}{3} \longleftarrow
$$

$$
3\overline{)14}
$$

$$
\underline{-12}
$$

$$
2 \longrightarrow
$$

$$
\frac{14}{3} = 4\frac{2}{3}
$$

To change a mixed number into an improper fraction, multiply the denominator of the fractional part by the whole number, and then add the product to the numerator of the fraction. The result is the numerator of the improper fraction, and the denominator remains the same as the fractional part of the mixed number.

Example: Write $2\frac{3}{4}$ as an improper fraction.

$$
2\frac{3}{4} = \frac{4 \times 2 + 3}{4} = \frac{11}{4}
$$

$$
2\frac{3}{4} = \frac{11}{4}
$$

PRACTICE PROBLEMS

1.5 Write $\frac{25}{4}$ as a mixed number.

1.6 Write $5\frac{7}{8}$ as an improper fraction.

Adding and Subtracting Fractions

To find the *sum*, means you add, and to find the *difference*, means you subtract. In order to add or subtract fractions, the fractions must have the same denominator, called a *common denominator*. If the denominators are not the same, you must find equivalent fractions that have the same denominator. To add or subtract the fractions with the common denominator, add or subtract the numerators only and place the result as the numerator of a fraction with the same denominator.

Example: Find the sum. $\frac{3}{4} + \frac{1}{10}$

Solution: Find equivalent fractions that have the same denominator. Add the numerators and use the same denominator.

Both 4 and 10 will divide into 20.

$$\frac{3}{4} = \frac{3 \times 5}{4 \times 5} = \frac{15}{20}$$

$$\frac{1}{10} = \frac{1 \times 2}{10 \times 2} = \frac{2}{20}$$

$$\frac{3}{4} + \frac{1}{10} = \frac{15}{20} + \frac{2}{20} = \frac{15 + 2}{20} = \frac{17}{20}$$

$$\frac{3}{4} + \frac{1}{10} = \frac{17}{20}$$

To add mixed numbers, add the fractions and then add the whole numbers. When you add the fractions, you may end up with an improper fraction. If so, change the improper fraction to a mixed number and add the whole number portions.

Example: Add. $2\frac{3}{7} + 4\frac{2}{3}$

Solution: Find a common denominator for the fractions and find the equivalent fractions. Add the fractions and then add the whole numbers. Simplify the answer if necessary.

Both 7 and 3 will divide into 21.

$$2\frac{3}{7} = 2\frac{3 \times 3}{7 \times 3} = 2\frac{9}{21}$$

$$4\frac{2}{3} = 4\frac{2 \times 7}{3 \times 7} = 4\frac{14}{21}$$

$$2\frac{3}{7} + 4\frac{2}{3} = 2\frac{9}{21} + 4\frac{14}{21} = 6\frac{9 + 14}{21} = 6\frac{23}{21}$$

$\frac{23}{21}$ is an improper fraction.

$$\frac{23}{21} = 1\frac{2}{21}$$

$$6 + 1\frac{2}{21} = 7\frac{2}{21}$$

$$2\frac{3}{7} + 4\frac{2}{3} = 7\frac{2}{21}$$

Subtracting mixed numbers will require borrowing if the numerator that is subtracted is larger than the numerator that it is subtracted from.

Example: Find the difference. $6\frac{1}{8} - 3\frac{5}{6}$

Solution: Find a common denominator for the fractions and find the equivalent fractions. Subtract the fractions, borrowing from the whole number if necessary. Subtract the whole numbers. Simplify the answer if necessary.

Both 8 and 6 will divide into 24.

$$6\frac{1}{8} = 6\frac{1 \times 3}{8 \times 3} = 6\frac{3}{24}$$

$$3\frac{5}{6} = 3\frac{5 \times 4}{6 \times 4} = 3\frac{20}{24}$$

Notice the numerators: 20 is larger than 3, so you must borrow and rename so that the numerator is larger than 20.

$$6\frac{3}{24} = 5 + 1\frac{3}{24}$$

Write $1\frac{3}{24}$ as an improper fraction: $1\frac{3}{24} = \frac{24 \times 1 + 3}{24} = \frac{27}{24}$

$$6\frac{3}{24} = 5\frac{27}{24}$$

$$6\frac{1}{8} - 3\frac{5}{6} = 5\frac{27}{24} - 3\frac{20}{24} = 2\frac{7}{24}$$

$$6\frac{1}{8} - 3\frac{5}{6} = 2\frac{7}{24}$$

PRACTICE PROBLEMS

1.7 $\quad 3\frac{1}{4} + 2\frac{2}{3}$

1.8 $\quad 8\frac{6}{7} + 1\frac{5}{14}$

1.9 $\quad 3\frac{5}{8} - 2\frac{1}{6}$

1.10 $\quad 6\frac{1}{3} - 5\frac{11}{12}$

Multiplying and Dividing Fractions

In a multiplication problem, numbers, called *factors*, are multiplied to find the *product*. To multiply fractions, multiply the numerators to find the numerator of the product and multiply the denominators to find the denominator of the product. Simplify the resulting fraction, if necessary.

Example: $\frac{2}{3} \times \frac{5}{8}$

Solution: Multiply the numerators, 2 and 5. Multiply the denominators, 3 and 8. Simplify if necessary.

$$\frac{2}{3} \times \frac{5}{8} = \frac{2 \times 5}{3 \times 8} = \frac{10}{24}$$

2 divides into the numerator and denominator.

$$\frac{10}{24} = \frac{10 \div 2}{24 \div 2} = \frac{5}{12}$$

$$\frac{2}{3} \times \frac{5}{8} = \frac{5}{12}$$

To divide by a fraction, multiply by the *reciprocal* of the fraction. To find the reciprocal of a fraction, invert the fraction, exchanging the numerator and denominator. For example, the reciprocal of $\frac{2}{3}$ is $\frac{3}{2}$.

Example: $\frac{6}{7} \div \frac{3}{4}$

Solution: Multiply $\frac{6}{7}$ by the reciprocal of $\frac{3}{4}$. Simplify if necessary.

$$\text{Reciprocal of } \frac{3}{4} = \frac{4}{3}$$

$$\frac{6}{7} \div \frac{3}{4} = \frac{6}{7} \times \frac{4}{3} = \frac{6 \times 4}{7 \times 3} = \frac{24}{21}$$

$$\frac{24}{21} = 1\frac{3}{21}$$

$$\frac{6}{7} \div \frac{3}{4} = 1\frac{3}{21}$$

To multiply or divide mixed numbers, change the mixed numbers to improper fractions and multiply or divide as in the previous examples.

Example: $1\frac{4}{5} \div \frac{3}{8}$

Solution: Change $1\frac{4}{5}$ to an improper fraction and multiply by the reciprocal of $\frac{3}{8}$.

$$1\frac{4}{5} = \frac{5 \times 1 + 4}{5} = \frac{9}{5}$$

$$\text{Reciprocal of } \frac{3}{8} = \frac{8}{3}$$

$$1\frac{4}{5} \div \frac{3}{8} = \frac{9}{5} \times \frac{8}{3} = \frac{9 \times 8}{5 \times 3} = \frac{72}{15}$$

$$\frac{72}{15} = 4\frac{12}{15}$$

3 divides into 12 and 15.

$$4\frac{12}{15} = 4\frac{12 \div 3}{15 \div 3} = 4\frac{4}{5}$$

$$1\frac{4}{5} \div \frac{3}{8} = 4\frac{4}{5}$$

PRACTICE PROBLEMS

1.11 $\frac{4}{5} \times \frac{2}{3}$

1.12 $\frac{3}{8} \div \frac{9}{16}$

1.13 $2\frac{3}{8} \div \frac{4}{5}$

1.2 Decimals

In a decimal number, each digit has a place value that is ten times the value of the digit to the right. The place value names are shown in Figure 1.1.

Hundred Thousands	Ten Thousands	Thousands	Hundreds	Tens	Ones	Decimal Point	Tenths	Hundredths	Thousandths	Ten Thousandths
4	8	2,	7	6	1	.	5	3	9	4

Figure 1.1
Place values.

The number shown in Figure 1.1 is read four hundred and eighty-two thousand, seven hundred sixty-one and five thousand three hundred ninety-four ten thousandths.

Since many numbers in business and personal finance are money amounts, many numbers will be rounded to the hundredths place, signifying hundredths of a dollar, or cents.

Rounding

To round a decimal, look to the digit to the right of the place value you are rounding to. If the digit to the right is 4 or below, keep the digit in the specified place the same. If the digit to the right is 5 or above, increase the digit in the specified place by one. All digits to the right of the place value you are rounding to become zero.

Example: Round 429,536 to the nearest thousand.

Solution: Find the thousands place. 42**9**,536

9 is in the thousands place.

The digit to the right of the 9 is 5.

Since 5 is 5 or greater, the thousands increases by 1.

9 becomes 10, or 429 thousand becomes 430 thousand.

All digits to the right of the 9 become 0.

42**9**,536 rounded to the nearest thousand is 430,000.

When you are rounding a number to a place to the right of the decimal point, you should drop the zeros to the right of the place you are rounding to.

Example: Round 263.7241 to the nearest hundredth.

Solution: Find the hundredths place. 263.7**2**41

2 is in the hundredths place.

The digit to the right of the 2 is 4.

Since 4 is less than 5, the hundredths place remains the same.

263.7241 rounded to the nearest hundredth is 263.7200

You should drop the zeros to the right of the hundredths place.

263.7241 rounded to the nearest hundredth is 263.7200 = 263.72.

PRACTICE PROBLEMS

1.14 Round to the nearest hundred: 829,467.22

1.15 Round to the nearest hundredth: 4,389.295

Operations with Decimals

Although most of the time you may be using a calculator to do calculations, it is still important to know how to do the operations without a calculator.

To add or subtract decimals, write the digits so the place values and the decimal point line up. You can write placeholder 0s in the place values to the right of the decimal point to assist in lining up the place values.

Example: Find the sum. $6.38 + 2.5 + 0.38$

Solution: Write 6.38, 2.5, and 0.38 so that the decimal points are aligned. Add a placeholder zero to 2.5 so that all numbers have the same number of digits after the decimal point. Add down the columns from right to left, carrying to the next columns the tens digit of each sum that is greater than 9. Bring the decimal point straight down.

$$
\begin{array}{r}
{\scriptstyle 1\ 1} \\
6.38 \\
2.50 \\
+\,0.38 \\
\hline
9.26
\end{array}
$$

$6.38 + 2.5 + 0.38 = 9.26.$

Example: Find the difference. $36 - 4.26$

Solution: Remember that the decimal point is to the right of a whole number. Align 36 and 4.26 vertically so the decimal points align. Add two placeholder zeros after the decimal point to 36 so that the numbers have the same number of digits after the decimal point. Subtract from right to left, borrowing as needed. Bring the decimal point straight down.

$$
\begin{array}{r}
{\scriptstyle 5\ \overset{9}{\cancel{10}}\ {}^{10}} \\
3\cancel{6}.\cancel{0}\cancel{0} \\
-\,4.26 \\
\hline
31.74
\end{array}
$$

$36 - 4.26 = 31.74.$

PRACTICE PROBLEMS

1.16 Find the sum. $2.857 + 0.46 + 0.2$

1.17 Find the difference. $28.2 - 16.35$

To multiply decimals, find each of the partial products and then add. Count the total number of decimal places in the factors being multiplied. Place the decimal in the product so that it has the same number of decimal places as the total number of decimal places in the factors.

Example: Find the product of 3.876 and 2.4.

Solution: Align the numbers so that the last digits of each are lined up. Multiply 4 by each of the digits of 3.876, from right to left. Drop down one line and write a placeholder zero under the last digit of the previous result. Multiply 2 by each of the digits of 3.876 from right to left. Add the two partial products. 3.876 has 3 decimal places, and 2.4 has one decimal place, so the product has four decimal places.

$$
\begin{array}{r}
{}^{1}3\,{}^{1}3\,{}^{1}2 \\
3.876 \\
\times \quad 2.4 \\
\hline
{}^{1}\;\;{}^{1} \\
15504 \\
+\;\;77520 \\
\hline
9.3024
\end{array}
$$

The product of 3.876 and 2.4 is 9.3024.

If the product does not have enough digits for the number of decimal places needed, add placeholder zeros before the leftmost digit.

Example: Find the product of 5.12 and 0.006.

Solution: Multiply 6 by the digits of 5.12. The product has five decimal places since 5.12 has two decimal places and 0.006 has three decimal places.

$$
\begin{array}{r}
{}^{1} \\
5.12 \\
\times\;0.006 \\
\hline
0.03072
\end{array}
$$

A zero is placed before the 3 to make five decimal places.

The product of 5.12 and 0.006 is 0.03072.

In a division problem, a *dividend* is divided by a *divisor* to find a *quotient*.

To divide decimals using long division, first move the decimal point in the divisor to the right so that it becomes a whole number. Move the decimal point in the dividend the same number of places, adding placeholder zeros if necessary. Place the decimal point in the quotient directly above the new position of the decimal point in the dividend.

Example: Find the quotient. 180.81 ÷ 6.3

Solution: Move the decimal point one place to the right to turn 6.3 into 63. Move the decimal point in 180.81 one place to the right to make 1808.1. Place the decimal point in the quotient directly above the decimal point in 1808.1. Divide.

$$
\begin{array}{r}
6.3\overline{)180.81} \\
\rightarrow\quad\rightarrow
\end{array}
\qquad
\begin{array}{r}
28.7 \\
6.3\overline{)180.81} \\
-\ 126\downarrow \\
\hline
54\boxed{8} \\
-\ 504\downarrow \\
\hline
44\boxed{1} \\
-\ 441 \\
\hline
0
\end{array}
$$

The quotient of 180.81 and 6.3 is 28.7.

PRACTICE PROBLEMS

1.18 Find the product of 3.7 and 0.003.

1.19 Divide. 3.647 ÷ 0.07

Fractions and Decimals

To change a fraction to a decimal, divide the numerator by the denominator.

Example: Change $\frac{3}{8}$ to a decimal.

Solution: Divide 3 by 8.

$$\begin{array}{r} 0.375 \\ 8\overline{)3.000} \\ -\underline{24} \\ 60 \\ -\underline{56} \\ 40 \\ -\underline{40} \\ 0 \end{array}$$

$$\frac{3}{8} = 3 \div 8 = 0.375$$

$$\frac{3}{8} = 0.375$$

Notice that when you calculate $3 \div 8$, you must add placeholder zeros after the decimal point in the dividend, or the 3.

To change a decimal to a fraction, determine the place value of the last digit of the decimal number. Put the digits to the right of the decimal as the numerator of a fraction and the number representing the place value of the decimal in the denominator. Reduce the fraction to lowest terms if necessary.

Example: Change 0.35 to a fraction.

Solution: 0.35 is 35 *hundredths*.

35 becomes the numerator.

100 becomes the denominator.

$$0.35 = \frac{35}{100}$$

5 divides into 35 and 100.

$$\frac{35}{100} = \frac{35 \div 5}{100 \div 5} = \frac{7}{20}$$

$$0.35 = \frac{35}{100} = \frac{7}{20}$$

PRACTICE PROBLEMS

1.20 Change $\frac{2}{5}$ to a decimal.

1.21 Change $\frac{4}{11}$ to a decimal rounded to the nearest hundredth.

1.22 Change 0.475 to a fraction.

1.3 Percents

Percent means "out of 100." A percent is a part out of 100. So, 25% means 25 out of 100, or $\frac{25}{100}$ or 0.25.

Percents are commonly used in business and personal finance. For example, sales tax may be 5% of the purchase price, or an account may earn 2% interest. Being able to change between fractions, decimals, and percent is an important part of mastering business and finance mathematics.

Fractions, Decimals, and Percents

To change a percent to a decimal, you can use the definition of percent— out of 100—to make a fraction. Place the number in front of the percent sign as the numerator and 100 as the denominator. Then change the fraction to a decimal.

Example: Change 56% to a decimal.

Solution: Change the percent to a fraction with a denominator of 100, and then change to a decimal.

$$56\% = 56 \text{ out of } 100 = \frac{56}{100}$$

$$\frac{56}{100} = 56 \div 100 = 0.56$$

$$56\% = 0.56$$

Since you are dividing by 100 to change the fraction to a decimal, you can use a shortcut to change a percent to a decimal.

> **TIP**
>
> To change a percent to a decimal, drop the percent sign and move the decimal point two places to the left.
>
> $$65\% = .65 = 0.65 \qquad 3\% = .03 = 0.03$$

To change a decimal to a percent, you can reverse the shortcut. Move the decimal point two places to the right, add any necessary zeros as placeholders, and attach the percent sign.

> **TIP**
>
> To change a decimal to a percent, move the decimal point two places to the right and add a percent sign.
>
> $$0.32 = 32 = 32\% \qquad 0.02 = 2 = 2\%$$

Example: Change 0.2 to a percent.

Solution: Move the decimal point two places to the right. Add a placeholder zero after the 2 to create two decimal places.

$$0.2 = 20 = 20\%$$

$$0.2 = 20\%$$

To change a fraction to a percent, first change the fraction to a decimal, and then move the decimal point two places to the right.

Example: Change $\frac{3}{5}$ to a percent.

Solution: Divide 3 by 5 to find the decimal equivalent. Move the decimal point two places to the right.

$$3 \div 5 = 0.60 = 60\%$$

$$\frac{3}{5} = 60\%$$

To change a percent to a fraction, use the definition of percent—out of 100—to make a fraction. Place the number in front of the percent sign as the numerator and 100 as the denominator. Reduce the fraction to lowest terms.

Example: Change 55% to a fraction.

Solution: Change 55% to a fraction out of 100, and reduce the fraction.

$$55\% = \frac{55}{100}$$

5 divides into 55 and 100.

$$\frac{55 \div 5}{100 \div 5} = \frac{11}{20}$$

$$55\% = \frac{11}{20}$$

PRACTICE PROBLEMS

1.23 Change 23% to a decimal.

1.24 Change 120% to a decimal.

1.25 Change 0.35 to a percent.

1.26 Change $\frac{4}{5}$ to a percent.

1.27 Change 28% to a fraction.

Operations with Percents

The most common application of percents in business and finance is to find the percent of a number. For example, sales tax may be 5% of the total sale. In mathematics, the word "of" typically indicates multiplication. To find a percent of an amount, multiply the decimal form of the percent by the amount.

Example: Find 18% of 125.

Solution: Change 18% to a decimal and multiply by 125.

$18\% = 0.18$

$125 \times 0.18 = 22.5$

22.5 is 18% of 125.

PRACTICE PROBLEMS

1.28 What is 22% of 75?

1.29 What is 4% of 125?

Another application problem using percents is to find the percent rate. For example, you can find the percent discount if a $15 item is discounted $6. A percent is represented by a part out of a whole of 100. To find the percent, write a fraction so that the numerator represents the part and the denominator represents the whole, and then change the fraction to a percent.

$$\frac{\text{Part}}{\text{Whole}} = \frac{6}{15} = 6 \div 15 = 0.40 = 40\%$$

The item is discounted 40%.

Example: 28 is what percent of 140?

Solution: Write a fraction with the $\frac{\text{part}}{\text{whole}}$. Divide to change the fraction to a decimal, and then move the decimal point two places to the right to change to a percent.

Fraction: $\dfrac{\text{part}}{\text{whole}} = \dfrac{28}{140}$

Change to decimal: $\dfrac{28}{140} = 28 \div 140 = 0.2$

Change to percent: $0.2 = 0.20 = 20\%$

28 is 20% of 140.

PRACTICE PROBLEMS

1.30 16 is what percent of 64?

1.31 22 is what percent of 140? Round to the nearest tenth of a percent.

1.4 Formulas

A *formula* is a rule for showing the relationships among variables. When you know the values for all of the variables but one, you can substitute the values into the formula and simplify to find the missing value.

Example: The formula to calculate simple interest is $I = P \times R \times T$, where I is the amount of interest, P is the principal, or amount invested, R is the interest rate, and T is the time. Find the amount of simple interest earned on $5,000 invested for six years at an 8% interest rate.

Solution: Identify the variables you have values for. Substitute the numbers for the variables and simplify.

$P = \$5,000, R = 8\% = 0.08, T = 6, I$ is unknown

$I = P \times R \times T$

$I = \$5,000 \times 0.08 \times 6$

$I = \$2,400$

The interest earned is $2,400.

PRACTICE PROBLEMS

1.32 Find G if $G = H \times R$ and $H = 45$ and $R = \$10.25$.

1.33 Find I if $I = P \times R \times T$ and $P = \$2,500, R = 2\%$ and $T = 3$ years

Exponents

Some formulas used in business and finance utilize *exponents*. An exponent is a mathematical notation used to indicate how many times a number, the base, is multiplied by itself.

Example: Find 3^4.

Solution: 3 is the base and 4 is the exponent.

Multiply the base 4 times.

$3 \times 3 \times 3 \times 3 = 81$

$3^4 = 81$

Most scientific calculators have an exponent key, typically an x^y or y^x key that you can use to calculate exponents quickly.

Order of Operations

To evaluate formulas that have multiple operations, you must apply the correct order of operations. For example, depending on what order you complete operations in, $2 + 3 \times 5$ might equal 25 or 17.

Follow these steps to do mathematical operations in the correct order:

1. Perform all operations inside parentheses.
2. Evaluate exponents.
3. Perform all multiplication and division from left to right.
4. Perform all addition and subtraction from left to right.

Example: Simplify $2 + 3 \times 5$.

Solution: There are no parentheses or exponents, so do the multiplication and then the addition.

$2 + 3 \times 5 = 2 + 15$	Perform multiplication.
$2 + 15 = 17$	Perform addition.
$2 + 3 \times 5 = 17$	

Example: Evaluate the formula $A = P(1 + r)^n$ if $P = 2,000$, $r = 0.02$, and $n = 5$. Round to the nearest hundredth.

Solution: Substitute the values into the formula. Evaluate to find A by using the order of operations.

$A = P(1 + r)^n$

$A = 2,000(1 + 0.02)^5$ Substitute values.

$A = 2,000(1.02)^5$ Perform operation inside parentheses.

$A = 2,000 \times 1.104080803$ Evaluate exponent.

$A = 2208.16$ Multiply and round to the nearest hundredth.

TIP

Wait until the end of a problem to round. When using a calculator, do not clear the calculator from one step to the next; instead, use the answer from one calculation to do the next calculation.

PRACTICE PROBLEMS

1.34 Evaluate the formula $A = P(1 + r)^n$ if $P = 1,500$, $r = 0.015$, and $n = 6$. Round to the nearest hundredth.

1.35 Find the APY if $APY = (1 + r)^n - 1$ and $r = 0.0125$ and $n \times 10$. Round to the nearest hundredth.

Chapter

2

Gross Pay

*G**ross pay*** is the amount of money an employee earns. Other names for gross pay include *gross wages* or *total earnings*. Typically, an employee receives less than the gross pay due to taxes and other deductions from their pay.

2.1 Hourly Pay

Employees who receive pay based on the number of hours that they work are paid an *hourly rate*. To find the gross pay for an hourly employee, multiply the number of hours that the employee worked by the hourly rate.

Gross Pay = Number of Hours Worked × Hourly Rate

Example: Jane works as a nurse and is paid $12.50 per hour. Last week, Jane worked 9 hours per day for 4 days. What is Jane's gross pay?

Solution: Multiply the number of hours worked per day by the number of days worked to find the number of hours worked. Multiply the number of hours worked by the hourly rate.

Hours worked = 9 hours per day × 4 days = 36 hours

Gross Pay = Number of Hours Worked × Hourly Rate

Gross Pay = 36 × $12.50 = $450

Jane's gross pay is $450.

PRACTICE PROBLEMS

2.1   Sydney worked 8 hours a day for 5 days. What is Sydney's gross pay if her hourly rate is $16.85 per hour?

2.2   Jared gets paid $6.75 per hour. Last week he worked the following hours: Monday $9\frac{1}{2}$ hours, Tuesday 5 hours, Wednesday $8\frac{3}{4}$ hours, Friday 10 hours. What is Jared's gross pay?

Overtime Wages

Overtime pay is earned by some employees for hours worked beyond the regular workday or workweek. For example, a worker may be paid overtime for any hours worked beyond 40 hours in a week or 8 hours in a day.

Often, an employer will pay *time-and-a-half rate* for overtime, although some employers pay *double-time rate* for overtime or working on a holiday.

* Time-and-a-Half Rate = 1.5 × Regular Pay Rate
* Double-Time Rate = 2 × Regular Pay Rate

TIP

Do not round overtime rates of pay.

Example: Chelsy's hourly rate is $8.25. Her employer will pay time-and-a-half if she works more than 8 hours in a day, and will pay double-time if she works on a holiday. What are her time-and-a-half and double-time rates of pay?

Solution: Multiply 1.5 by the regular pay rate to find the time-and-a-half rate and multiply 2 by the regular pay rate to find the double-time rate.

Time-and-a-Half Rate $= 1.5 \times$ Regular Pay Rate

Time-and-a-Half Rate $= 1.5 \times \$8.25 = \$12.375$ (Do not round.)

Double-Time Rate $= 2 \times$ Regular Pay Rate

Double-Time Rate $= 2 \times \$8.25 = \$16.50$

Chelsy's overtime rates of pay are $12.375 for time-and-a-half and $16.50 for double-time.

To find gross pay when there are regular and overtime hours, calculate the gross pay for the regular hours and the gross pay for the overtime hours, and then add them together.

Example: Josef is paid $10.75 per hour for regular time, and time-and-a-half for hours beyond 40 hours per week. If Josef worked 8 hours, 12 hours, 7 hours, 10 hours, and 9 hours in one week, what is his gross pay?

Solution: Add the hours to find the total hours worked. Subtract 40 from the number of hours worked to determine how many hours are overtime hours. Calculate the regular wages, the overtime rate of pay, and the overtime wages. Add the overtime and regular wages.

Total hours worked $= 8 + 12 + 7 + 10 + 9 = 46$

Overtime hours $= 46 - 40 = 6$

Josef worked 40 regular hours and 6 overtime hours.

Regular pay $= 40 \times \$10.75 = \$430$

Overtime rate of pay $= \$10.75 \times 1.5 = \$16.125$ (Do not round.)

Overtime pay $= 6 \times \$16.125 = \$96.75$

Gross Pay $= \$430 + \$96.75 = \$526.75$

Josef's gross pay is $526.75.

PRACTICE PROBLEMS

2.3 Clint earns $25.50 per hour as a mechanic. He is paid time-and-a-half for working more than 8 hours per day. Yesterday, he worked $11\frac{1}{2}$ hours. What is his gross pay for the day?

2.4 Yvette worked 50 hours last week. She earns time-and-half for any hours over 40 and up to 45 hours in a week and double-time for any hours over 45. If her regular pay is $10.75 per hour, what are her gross wages for last week?

2.2 Salary

A *salary* is a fixed amount of pay for each pay period. A salaried employee does not receive overtime pay. In addition, a salaried employee gets paid the same amount even if the employee uses sick days or vacation days.

Salaries are often expressed as a yearly amount, but the salary is paid out monthly, bi-monthly, or semi-monthly, bi-weekly, or weekly.

• Monthly: Once a month, 12 times a year

• Bi-monthly or semi-monthly: Twice a month, 24 times a year

• Bi-weekly: Every other week, 26 times a year

• Weekly: Once a week, 52 times a year

Notice the different meaning of the prefix *bi-* in the terms bi-monthly and bi-weekly. The prefix bi- can mean either "occurring twice" or "every other," and in the business world, bi-monthly pay means pay periods twice a month, while bi-weekly pay means a pay period every other week.

To find the gross pay per pay period, divide the yearly salary by the number of pay periods.

Example: Willa earns a salary of $33,000 per year. She is paid monthly. What is her gross pay per month?

Solution: There are 12 months in a year. Divide the salary by 12 to find the monthly pay.

$33,000 ÷ 12 = $2,750

Willa's gross pay is $2,750 per month.

To find the gross yearly pay when you know the pay per pay period, multiply the pay per pay period by the number of pay periods in a year.

Example: Juan earns a salary of $350 per week. What is his yearly salary?

Solution: There are 52 weeks in a year. Multiply the weekly salary by 52 to find the yearly salary.

$350 × 52 = $18,200

Juan earns $18,200 per year.

Practice Problems

2.5 Melissa earns a salary of $2,000 per month, paid semi-monthly. What is her gross pay each pay period?

2.6 Max is offered a job that pays $36,000 per year. He estimates that he needs monthly gross wages of $3,250 to cover his living expenses. Does the job pay enough?

2.7 Sydney is considering two jobs. One job pays $45,000 per year. The other job pays a weekly salary of $950. Which job pays better?

2.3 Commission

Often salespeople will be paid a *commission*, an amount for every item sold or a percentage of the value of their sales. Some employees earn only commission, called *straight commission*, while others earn a salary or hourly rate plus a commission.

If the employee is paid a fixed commission for each item sold, multiply the quantity sold by the rate of commission.

Commission = Quantity Sold × Rate of Commission

Example: Jillian is paid a commission of $48.50 for each original design she creates and sells. If she creates and sells 36 designs, how much will she earn?

Solution: Multiply the quantity sold by the rate of commission.

Commission = Quantity Sold × Rate of Commission

Commission = 36 × $48.50 = $1,746

Jillian earns $1,746.

If the employee is paid a percentage of the value of her sales, multiply the amount of sales by the rate of commission. The rate of commission will usually be expressed as a percent. Change the percent to a decimal number.

Commission = Sales × Rate of Commission

Example: A car salesman earns 5% on his sales. In one month, his sales totaled $120,000. How much did the car salesman earn?

Solution: Change the rate of commission to a decimal. Multiply the sales by the decimal form of the rate of commission.

Rate of Commission = 5% = 0.05

Commission = Sales × Rate of Commission

Commission = $120,000 × 0.05 = $6,000

The car salesman earns $6,000 in commission.

PRACTICE PROBLEMS

2.8   A door-to-door salesman earns a commission of $8.50 for each encyclopedia set he sells. If he sells 50 sets in a week, how much will he earn in commission?

2.9   Joshua is a real estate agent and he earns a 3% commission on each home sale. If he sells a home for $325,000, how much does he earn in commission?

Commission Based on Quota

Some employers will pay a commission on any sales an employee makes above a certain level, or *quota*. Often these employees will earn a salary or hourly rate in addition to the commission.

To calculate the commission, subtract the quota from the sales and multiply by the rate of commission. The gross pay is the salary or hourly pay plus the commission.

- Commission = (Sales − Quota) × Rate of Commission
- Gross Pay = Salary + Commission
- Gross Pay = Hourly Pay + Commission

> **TIP**
>
> Use the order of operations to find the commission. Evaluate what is inside the parentheses first.

Example: Rashid works as a salesperson in an electronics shop. He earns a weekly salary of $250 plus a 3% commission on sales over $2,000 per week. If Rashid's sales in one week were $3,500, what are his gross wages for the week?

Solution: Subtract $2,000 from $3,500 to find the amount of sales on which he earns commission. Multiply by the decimal form of the commission rate. Add the commission to the weekly salary.

Rate of commission = 3% = 0.03

Commission = (Sales − Quota) × Rate of Commission

Commission = ($3,500 − $2,000) × 0.03

Commission = $1,500 × 0.03 = $45

Gross Pay = Salary + Commission

Gross Pay = $250 + $45 = $295

Rashid earns $295 for the week.

PRACTICE PROBLEMS

2.10 Sai-Ling earns a 2% commission on sales above $10,000 and a monthly salary of $1,200. If her sales last month were $55,000, what were her total earnings for the month?

2.11 Edward is paid a 10% commission on sales above $75,000 and an hourly rate of $6.50 per hour. If his sales were $125,000 last week and he worked 40 hours, what were his gross wages?

Graduated Commission

An employee who is paid a *graduated commission* is paid an increasing rate of commission as the amount of sales increases.

Example: Connie works as a salesperson. She earns 5% commission on the first $10,000 of sales, and 8% on any sales above $10,000 for the week. How much did Connie earn in commission in a week if her sales were $25,000?

Solution: Calculate the commission for the sales up to $10,000 and the commission for the sales over $10,000. Add the two commissions together to find the total commission.

Rate of commission up to $10,000 = 5% = 0.05

Commission on first $10,000 = $10,000 × 0.05 = $500

Rate of commission for over $10,000 = 8% = 0.08

Amount of sales over $10,000 = $25,000 − $10,000 = $15,000

Commission on sales over $10,000 = $15,000 × 0.08 = $1,200

Total commission = $500 + $1,200 = $1,700

Connie earned $1,700 in commission.

PRACTICE PROBLEMS

2.12 Jeremiah earns a $2.50 commission on the first 30 prints he sells at an art show. He earns $3.25 for each print he sells after the first 30. In one weekend, he sold 100 prints at an art show. How much did he make in commission?

2.13 A company pays salespeople a commission of 4% on the first $20,000 of sales, 8% on sales over $20,000, and up to $40,000, and 10% commission on sales above $40,000. The highest-selling salesperson had $150,000 in sales. What commission did the salesperson earn?

Rate of Commission

If the amount of sales and the amount of commission is known, you can find the rate of commission. Divide the amount of commission by the sales to find a decimal number. Change the decimal to a percent.

$$\text{Rate of Commission} = \frac{\text{Amount of Commission}}{\text{Sales}}$$

Example: Julio started a new job, and the first week, he earned $315 in straight commission. If his sales for the week were $4,500, what is his rate of commission?

Solution: Divide the commission by the sales and change the decimal to a percent.

$$\text{Rate of Commission} = \frac{\text{Amount of Commission}}{\text{Sales}}$$

$$\text{Rate of Commission} = \frac{\$315}{\$4,500} = 0.07 = 7\%$$

Julio's rate of commission is 7%.

PRACTICE PROBLEMS

2.14 A worker earns $1.12 for each program he sells. The programs cost $8. What percent commission does the worker earn?

2.15 Corina works at a boutique and earns $9.75 per hour plus commission. Last month she worked 150 hours and had $5,000 in sales. Her gross pay was $1,712.50. What is her rate of commission?

2.4 Other Wage Plans

There are many different kinds of wage plans in addition to hourly rates, salary, and commission. Employees might be paid based on each item they produce. They might be paid a fixed amount per day, or receive tips in addition to their regular pay.

Piece-Rate Pay

Piece-rate employees are paid for the number of pieces they produce that meet the employer's standard. To find the gross pay, multiply the number of pieces produced by the piece rate.

$$\text{Gross Pay} = \text{Number of Pieces Produced} \times \text{Piece Rate}$$

Example: Frank earns $3.25 for each quality necklace he produces. If he produces 40 on Monday, 35 on Tuesday, 55 on Wednesday, and 68 on Thursday, how much will he earn?

Solution: Find the total number of necklaces produced and multiply by the piece rate.

Necklaces produced = 40 + 35 + 55 + 68 = 198

Gross Pay = Number of Pieces Produced × Piece Rate

Gross Pay = 198 × $3.25 = $643.50

Frank earns $643.50.

PRACTICE PROBLEM

2.16 An artist is hired to produce tiles commemorating a local historic event that will be sold at the county fair. The artist is paid $4.50 for each tile, which will be sold for $15 at the fair. If the artist produces 400 tiles, what is her gross pay?

Per Diem Pay

Per diem means by the day. Some employees, particularly temporary employees, may be paid wages on a per diem basis, or a fixed amount per day. To find the gross pay, multiply the number of days worked by the per diem rate.

$$\text{Gross Pay} = \text{Number of Days} \times \text{Per Diem Rate}$$

Example: Stella works as a temporary employee for a publishing company, earning $75 per day. If she worked 14 days last month, what is her gross pay?

Gross Pay = Number of Days × Per Diem Rate

Gross Pay = 14 × $75 = $1,050

Stella's gross pay is $1,050.

PRACTICE PROBLEM

2.17 Angelica works as a substitute teacher. She receives per diem pay of $80 per day. Last month, she worked 12 days as a substitute teacher. What is Angelica's gross pay?

Tip Pay

Some employees can earn tips, or *gratuity*. Tips are paid voluntarily by the people the employee serves. Often tips are paid as a percentage of the client's total bill. For example, a restaurant patron may pay the waitress a 20% tip.

Tip = Total Bill × Tip Percentage

Other employees may receive tips based on the amount of service provided. For example, a hotel guest may pay a bellhop $2 per bag for taking luggage to the hotel room.

Tip = Number of Units × Tip Rate

Example: A family eats at a restaurant, and they receive very good service from the waiter. The family wants to leave a 20% tip. How much will the tip be if the total bill is $56.25?

Solution: Change the tip percentage to a decimal. Multiply the total bill by the tip percentage.

Tip percentage = 20% = 0.20

Tip = Total Bill × Tip Percentage

Tip = $56.25 × 0.20 = $11.25

The tip is $11.25.

> **TIP**
>
> An easy way to calculate a 10% tip is to move the decimal point in the amount of the bill one place to the left. For example, for a bill of $56.25, moving the decimal point one place to the left makes a 10% tip of $5.625 or $5.63. For ease of calculation, you might round the tip amount to $5.50. For a 20% tip, double the 10% tip. If the 10% tip was rounded to $5.50, then a 20% tip would be about $5.50 $\times$ 2, or $11.00. To estimate a 15% tip, cut the 10% tip in half and add that amount back to the 10% tip. For the 10% tip of $5.50, $5.50 $\div$ 2 = $2.75, and $5.50 + $2.75 = $8.25, so $8.25 is about a 15% tip.

PRACTICE PROBLEMS

2.18 At an airport, the skycaps suggest a tip of $2.00 per bag for each piece of luggage that the skycap handles. A family of four is traveling, and each person has two bags. What is the suggested tip for handling all of the family's bags?

2.19 A client at a hair salon plans to leave a 15% tip. The client's total bill is $85. What tip should the client leave?

2.5 Average Pay

Many employees do not earn the same amount every day or every pay period due to commission, tips, or varying number of hours worked. An *average* is the best way to express how much they make.

Averages

To find the average, or *mean*, of a group of numbers, add the numbers together and divide by the number of items added.

Example: Joe earns an hourly rate of $4.05 per hour. In six months he worked the following hours: January 160 hours, February 200 hours, March 140 hours, April 175 hours, May 220 hours, and June 150 hours. What is Joe's average monthly gross pay?

Solution: Add the number of hours worked in 6 months and divide the sum by 6 to find the average number of hours worked per month. Multiply the average number of hours by the hourly rate to find the average monthly gross pay.

$$\text{Average hours} = \frac{160 + 200 + 140 + 175 + 220 + 150}{6}$$

$$= \frac{1{,}045}{6} = 174.1667 \text{ hours}$$

Average monthly gross pay = $174.1667 \times \$4.05 = \$705.38$

Joe's average monthly gross wage is $705.38.

TIP

If you are using a calculator to find an average, make sure you press the equals button to total the numbers before you divide. If you do not, the calculator will only divide the last number by the number of items.

Instead of a list of numbers, data may be presented in groups. It is important to account for all of the numbers before finding the average.

Example: Last week, Sondra worked the following hours: one day 7 hours, three days 10 hours, two days 11 hours, and one day 4 hours. What was the average number of hours she worked per day last week?

Solution: Find the total number of hours worked last week and divide by the number of days.

1 day @ 7 hours = $1 \times 7 = 7$ hours

3 days @ 10 hours = $3 \times 10 = 30$ hours

2 days @ 11 hours = $2 \times 11 = 22$ hours

1 day @ 4 hours = $1 \times 4 = 4$ hours

Total hours = $7 + 30 + 22 + 4 = 63$ hours worked

Average $= \frac{63}{7} = 9$ hours per day

Sondra worked an average of 9 hours per day.

PRACTICE PROBLEMS

2.20 In one week, a waiter earned the following in tips: $85, $48, $75, $125, $63. If his gross hourly wages for the 5 nights of work was $230, what were his average gross earnings per night?

2.21 In the last year, Shana has earned a commission each month. Two months she earned $2,500 in commission, three months she earned $4,000 in commission, six months she earned $3,000 in commission, and one month she earned $8,000 in commission. What was her average monthly commission?

Average as a Goal

A business may use an average, such as average monthly or yearly sales, as a goal or standard by which to evaluate employees or the business. You may need to calculate what amount is necessary in order to reach that average.

The average is calculated by dividing the sum of a group of items by the number of items. To "undo" the average, multiply the average by the number of items so that you can find the sum that will produce the desired average.

Example: A company prides itself that its office staff makes an average of at least $17 per hour. They currently employ three workers who make $15.75, $18.55, and $17.25 per hour. The company is adding a fourth position to the office staff. How much should the fourth person be paid in order for the average pay of the office staff to remain at least $17 per hour?

Solution: Multiply the average by 4 to find the total pay for four workers to meet the $17 average. Add the pay of the current workers and subtract from the total pay to find the amount the fourth person should be paid.

Total pay for four workers to meet average = $17 × 4 = $68

Total pay for current workers = $15.75 + $18.55 + $17.25 = $51.55

Pay needed for new worker = $68 – $51.55 = $16.45

You can check your work by finding the average pay of the four workers.

$$\text{Average} = \frac{\$15.75 + \$18.55 + \$17.25 + \$16.45}{4} = \frac{\$68}{4} = \$17$$

The fourth person should be paid at least $16.45.

PRACTICE PROBLEMS

2.22 In the last four months, Gary has earned a total of $15,000 in commission. How much does he need to earn in commission in the fifth month in order to have a monthly average of $5,500 in commission?

2.23 A company gives a bonus to all salespeople who have a monthly sales average of $100,000 or more. Madge has the following sales for the first 11 months of the year: two months $85,000, three months $125,000, four months $90,000, and two months $80,000. How much does she need to sell in the final month of the year in order to receive the bonus?

Chapter

3

Net Pay

While gross pay is the amount of pay that you earn, *net pay* is the amount of money that you are actually paid after taxes and other amounts are deducted from your pay.

Tax laws are complex and change regularly. The illustrations given in this chapter are only a brief introduction to basic tax calculations. Taxpayers should follow the instructions given in government publications or consult a lawyer, accountant, or tax professional when determining how tax issues apply to a specific situation.

3.1 Deductions from Pay

Deductions are amounts subtracted from gross pay. Federal income tax, or *withholding tax*, Social Security, Medicare, city and state income taxes, and health insurance premiums are common deductions.

Federal Withholding

For federal income taxes, the amount withheld from a worker's pay will vary depending on the wages, marital status, and the number of withholding allowances claimed. *Withholding allowances* reduce the amount of tax withheld. Workers can claim allowances for themselves, a spouse, and for children and other dependents.

Current federal withholding tables change from year to year and can be found in a variety of sources. A sample withholding table for married employees paid on a monthly basis is shown in Figure 3.1.

MARRIED Persons—**MONTHLY** Payroll Period

And the wages are—		And the number of withholding allowances claimed is—				
At least	But less than	0	1	2	3	4
		The amount of income tax to be withheld is—				
2280	2320	128	85	55	24	0
2320	2360	134	89	59	28	0
2360	2400	140	95	63	32	2
2400	2440	146	101	67	36	6
2440	2480	152	107	71	40	10
2480	2520	158	113	75	44	14
2520	2560	164	119	79	48	18
2560	2600	170	125	83	52	22
2600	2640	176	131	87	56	26
2640	2680	182	137	91	60	30
2680	2720	188	143	97	64	34

Figure 3.1
Sample federal withholding table.

Use the first two columns to find the row that has the correct wage bracket for the employee's wages. Identify the column that represents the number of withholding allowances claimed. The amount of tax to be deducted from the gross pay will be in that row and column.

Example: Luann works as a paralegal and is paid a monthly salary of $2,500. She is married and has one child. She claims 3 withholding allowances. How much federal income tax should be withheld?

Solution: Locate the wage bracket and number of withholding allowances in the table, and find the tax.

Luann makes at least $2,480, but less than $2,520. She claims 3 exemptions.

Luann will have $44 of federal income tax withheld from her monthly salary.

PRACTICE PROBLEMS

3.1 Jonisha earns a monthly salary of $2,555. She is married and claims 4 withholding allowances. How much is deducted from her pay for federal income tax?

3.2 Mark is married and currently claims two withholding allowances. His wife is about to have a baby, and he will change the number of withholding allowances to 3. If he earns a monthly salary of $2,450, how much will the amount withheld for federal income tax decrease?

Social Security and Medicare Tax

Social Security and Medicare taxes together are called *FICA* tax, since they are part of the Federal Insurance Contributions Act. Since 1990, the rate for Social Security tax has been 6.2% of wages up to a certain amount. The maximum wages subject to the Social Security tax has increased over time, with the maximum wage at $106,800 since 2009.

For the year 2011, the Social Security tax rate for employees was reduced to 4.2% on the first $106,800 of wages. The Social Security tax rate reduction for employees is scheduled to expire at the end of 2011. Unless the reduction is extended, the Social Security tax rate for employees will return to 6.2% beginning in 2012. The Medicare rate is 1.45% on all wages.

Both employees and employers pay FICA taxes on the employee's wages. The rate for the employer's Social Security tax remains at 6.2% during 2011. The employer also pays 1.45% on all employee wages for Medicare tax. The employee's share of FICA tax is deducted from the employee's gross pay. To find the FICA deduction from gross pay, multiply the wages by each tax rate, and add the taxes together. If the worker has not yet earned $106,800, you can multiply the wages by the combined FICA tax rate. For 2011, the combined FICA tax rate is 5.65% (4.2% + 1.45%). If the Social Security tax rate increases back to 6.2% after 2011, then the combined FICA tax rate will increase to 7.65% (6.2% + 1.45%).

Example: Denny has a gross weekly pay of $1,255. How much will be deducted from his pay each month for FICA tax using the 2011 rates?

Solution: Check to make sure the worker will not exceed the $106,800 wage limit. If not, change the combined FICA tax rate of 5.65% to a decimal and multiply the gross pay by the decimal tax rate.

At $1,255 per week, Denny will earn $1,255 × 52 = $65,260 per year, so he will not reach the $106,800 wage limit.

Combined FICA tax rate = 5.65% = 0.0565

FICA tax = $1,255 × 0.0565 = $70.91

$70.91 will be deducted for FICA tax.

Example: Janet has earned $105,900 so far this year. Her gross wages on her next paycheck are $4,250. How much will be deducted for Social Security and Medicare taxes using the 2011 rates?

Solution: Subtract the wages earned to date from the Social Security wage limit to find the amount of wages subject to Social Security tax. Multiply those wages by the Social Security tax rate. Multiply the gross wages by the Medicare tax rate. Add the Social Security and Medicare taxes together.

Wages subject to Social Security tax: $106,800 − $105,900 = $900

Social Security tax rate = 4.2% = 0.042

Social Security tax = $900 × 0.042 = $37.80

Medicare tax rate = 1.45% = 0.0145

Medicare taxes = $4,250 × 0.0145 = $61.63

FICA tax = $37.80 + $61.63 = $99.43

Janet will have $99.43 deducted for FICA tax.

PRACTICE PROBLEMS

3.3 Joan earns $10,800 per month. How much will she pay in Social Security and Medicare taxes for February 2011?

3.4 How much will Joan pay in Social Security and Medicare taxes in December 2011?

3.5 What is the maximum yearly amount any person will pay in Social Security taxes in 2011? What will the maximum yearly amount be if the Social Security tax rate returns to 6.2%?

Net Pay

Net pay is the amount of pay after all deductions. Net pay is also called *take home pay*. In addition to taxes, there may be other deductions from gross pay, such as health or life insurance premiums.

To find net pay, add all of the deductions and subtract from the gross pay.

$$\text{Net Pay} = \text{Gross Pay} - \text{Deductions}$$

Example: Jaime earned gross wages of $683. The following deductions were taken from her wages: health insurance $50, federal withholding $71, and Social Security and Medicare taxes $38.59. What is Jaime's net pay?

Solution: Add all of the deductions and subtract from the gross wages.

Total deductions = $50 + $71 + $38.59 = $159.59

Net pay = $683 − $159.59 = $523.41

Jaime's net pay is $523.41.

PRACTICE PROBLEMS

3.6 Gloria earned $955 last week. She had the following deductions: Social Security and Medicare taxes $53.96, federal withholding $78, life insurance $12, health insurance $45. What is Gloria's net pay?

3.7 Pam is married and claims 2 withholding allowances. She earns a monthly salary of $2,645. She has the following deductions: federal withholding, Social Security and Medicare taxes, union dues $35, health insurance $150. What is Pam's net pay? Use Figure 3.1 for federal withholding and 5.65% for FICA tax.

3.2 Federal Income Taxes

The United States federal income tax is a pay-as-you go system. Employers withhold money from employees' pay all year. After the end of the year, workers file an income tax return to determine whether too little or too much in federal taxes was withheld from their wages during the year. Taxpayers must pay more money if too little was withheld, or they are eligible for a refund if too much was withheld. Under some conditions, if too little was withheld, a taxpayer can be subject to penalties and interest.

Adjusted Gross Income and Taxable Income

For tax purposes, *gross income* is the total income for the year including wages, tips, bonuses, interest, and income from other sources. The government allows certain adjustments to income that decrease gross income, such as money paid into approved retirement plans. The result after these adjustments is called your *adjusted gross income*.

Adjusted Gross Income = Gross Income − Adjustments to Income

You can subtract from your adjusted gross income exemptions and deductions for which you qualify. The result is your *taxable income*, or the income for which you are responsible to pay taxes.

Taxable Income = Adjusted Gross Income −
(Deductions + Exemptions)

Example: Terry earned $28,455 last year in gross income. He has made $1,800 in payments into an approved retirement program. He has deductions of $5,700 and exemptions of $3,650. What is Terry's taxable income?

Solution: Subtract adjustments from the gross income to find the adjusted gross income. Add the deductions and exemptions and subtract from the adjusted gross income.

Adjusted gross income = $28,455 − $1,800 = $26,655

Total deductions and exemptions = $5,700 + $3,650 = $9,350

Taxable income = $26,655 − $9,350 = $17,305

Terry's taxable income is $17,305.

Practice Problems

3.8 Jeremy earned $65,405 last year. He qualifies for $2,500 in adjustments to income, $10,850 in deductions and exemptions of $3,650. What is Jeremy's taxable income?

3.9 Javier earned $22,000 last year. He has $500 in adjustments to income, $5,800 in deductions and exemptions of $3,650. What is Javier's taxable income?

Income Tax Due

If your taxable income is less than $100,000, you use a tax table to find the tax due. Tax tables change each year and are provided with federal tax forms. A portion of a sample federal income tax table is shown in Figure 3.2.

If line 43 (taxable income) is—		And you are—			
At least	But less than	Single	Married filing jointly *	Married filing separately	Head of a household
		Your tax is—			
26,000					
26,000	26,050	3,486	3,069	3,486	3,306
26,050	26,100	3,494	3,076	3,494	3,314
26,100	26,150	3,501	3,084	3,501	3,321
26,150	26,200	3,509	3,091	3,509	3,329
26,200	26,250	3,516	3,099	3,516	3,336
26,250	26,300	3,524	3,106	3,524	3,344
26,300	26,350	3,531	3,114	3,531	3,351
26,350	26,400	3,539	3,121	3,539	3,359

Figure 3.2
Sample federal income tax table.

Use the first column to find the row that has the correct wage bracket. Identify the column that represents the filing status. The amount of tax due will be in that row and column.

Example: Felicia has a taxable income of $26,155. If she is married filing jointly, how much federal income tax is due?

Solution: Find the row that represents her wage bracket, and then move to the column under Married Filing Jointly heading.

Felicia's taxable income is at least $26,150 but less than $26,200.

$3,091 in federal income tax is due.

Amount Due or Refund

Once you have found the amount of tax due, compare that to the total federal income tax that has been withheld. If more has been withheld than the tax due, you are eligible for a *refund* for the amount you over-paid. If less has been withheld than the amount of tax due, then you will need to calculate the amount due and pay more.

To find out the amount due or the refund, subtract the smaller number from the larger.

- Amount Due = Tax Due − Tax Withheld
- Refund = Tax Withheld − Tax Due

Example: Tang is married and will be filing his taxes jointly with his spouse. He has a taxable income of $26,025. His employer withheld $3,555 for federal income tax. How much income tax is due? Is Tang eligible for a refund, or is there an amount due? How much?

Solution: Use Figure 3.2 to find the income tax due. Compare the tax due to the amount that was withheld from Tang's wages. If the tax due is greater than the amount withheld, Tang needs to pay more. If the amount withheld is greater than the tax due, Tang is eligible for a refund. Subtract the smaller amount from the larger amount.

Tang's taxable income is at least $26,000, but less than $26,050.

The amount of income tax due is $3,069.

Tang's employer withheld $3,555.

$3,555 is larger than $3,069, so Tang is eligible for a refund.

Refund = Tax Withheld − Tax Due

Refund = $3,555 − $3,069 = $486

Tang is eligible for a refund of $486.

When a taxpayer is eligible for a refund on his federal income taxes, he has the option to receive the refund by check or direct deposit into a bank account, or he can have part or all of the refund applied to the following year's taxes.

PRACTICE PROBLEMS

3.10 Bonita is single and has a taxable income of $26,265. Her employer withheld $3,289 for federal income tax. How much income tax is due? Does Bonita need to pay more income tax, or is she eligible for a refund? How much?

3.11 Torrie files her taxes as a head of household. Her taxable income is $26,225. Her employer deducted $3,023 in withholding tax from her pay. Does Torrie need to pay more income tax, or is she eligible for a refund? How much?

3.3 State and City Income Taxes

Some cities and most states collect income taxes. These taxes are paid in addition to federal income taxes. Each city and state that has an income tax determines how much of taxpayer's income is taxable and the tax rate that is applied.

Flat Income Taxes

A *flat tax* is a tax that is charged using a single tax rate applied to all taxpayers. A handful of states have a flat income tax. Some states apply the flat tax rate to all income, while others apply the single tax rate to a taxpayer's federal adjusted gross income, federal taxable income, or some modification of the federal adjusted gross or taxable income.

Example: Jarrod's federal taxable income is $32,895. The state income tax rate is 4.35% of federal taxable income. What is Jarrod's state income tax due?

Solution: Change the tax rate to a decimal and multiply by the federal taxable income.

State tax rate = 4.35% = 0.0435

State income tax = $32,895 × 0.0435 = $1,430.93

Jarrod owes $1,430.93 in state income tax.

PRACTICE PROBLEMS

3.12 Ming's taxable income is $14,866. The state income tax rate is 3% of taxable income. How much does Ming owe in state income tax?

3.13 Jack lives in a state that taxes residents 4.8% of that taxpayer's federal adjusted gross income. Jack's gross income last year was $55,450. He has adjustments totaling $3,000. How much does Jack owe in state income tax?

Progressive Income Taxes

A *progressive*, or *graduated*, tax charges more in taxes as your income increases. The United States federal income tax is an example of a progressive tax. City and state progressive tax rates vary widely. A sample is shown in Table 3.1.

Table 3.1 Sample Progressive Tax Rate

Tax Rate	Taxable Income Range
2.59%	on the first $10,000 of taxable income
2.88%	on taxable income between $10,001 and $25,000
3.36%	on taxable income between $25,001 and $50,000
4.24%	on taxable income between $50,001 and $150,000
4.54%	on taxable income over $150,000

Example: Sai-Ling's taxable income last year was $30,000. If her state uses the tax rate shown in Table 3.1, what is her state income tax due?

Solution: Sai-Ling's income falls into the third income bracket of $25,001 to $50,000, but she pays the lowest tax rate of 2.59% on the first $10,000, 2.88% on the next $15,000 for the income that is between $10,001 and $25,000, and 3.36% on the remaining $5,000. Multiply the amount of income at each bracket by the tax rate to find the tax for that income. Add the taxes together to find the total tax due.

Tax on first $10,000 = $10,000 × 2.59% = $10,000 × 0.0259 = $259

Tax on $10,001 to $25,000 = $15,000 × 2.88%
$$= \$15,000 \times 0.0288 = \$432$$

Tax on last $5,000 = $5,000 \times 3.36\% = $5,000 \times 0.0336 = $168

Total tax due = $259 + $432 + $168 = $859

Sai-Ling's state income tax is $859.

PRACTICE PROBLEMS

3.14 Jorge's taxable income last year was $36,300. Using the progressive tax rates in Table 3.1, how much state income tax does Jorge owe?

3.15 Michelle has a taxable income of $50,000. Next year, she is expecting a raise, and she estimates her taxable income will increase to $52,000. How much more will she owe in state income tax if the progressive tax rate in Table 3.1 stays the same?

3.4 Employee Benefits and Expenses

Many employers offer *employee benefits*, or *fringe benefits* in addition to pay. Benefits might include things like health insurance, pension plan, sick leave, vacation time, or the use of a company car.

When considering an offer of employment, it is helpful to consider the *total employee benefits* in addition to the wages. In a job offer, the employee benefits may be expressed in dollars or as a percentage of the wages.

To find the total employee benefits, either add the value of the benefits stated in dollars or multiply the gross pay by the benefits percentage rate, expressed as a decimal.

- Total Employee Benefits = Sum of the Benefits
- Total Employee Benefits = Gross Pay × Benefit Rate

Total Job Benefits

To find the *total job benefits*, add the gross pay and the total employee benefits.

Total Job Benefits = Gross Pay + Total Employee Benefits

Example: John is considering two job offers. One job has a salary of $38,000 per year and a benefits package equal to 20% of his salary. The other job offers a monthly salary of $2,750 plus the following annual benefits: pension $2,000, paid vacation $1,250, health insurance $850. What are the total job benefits for each offer?

Solution: For the first job, change the benefits percentage rate to a decimal and multiply it by the salary to find the total employee benefits. For the second job, find the yearly salary by multiplying the monthly salary by 12. Add the value of the employee benefits to find the total employee benefits. Add the total employee benefits to the yearly salary for each job.

Job 1: Benefit rate = 20% = 0.20

Total Employee Benefits = $38,000 × 0.20 = $7,600

Total Job Benefits = $38,000 + $7,600 = $45,600

Job 2: Annual Salary = $2,750 × 12 = $33,000

Total Employee Benefits = $2,000 + $1,250 + $850 = $4,100

Total Job Benefits = $33,000 + $4,100 = $37,100

The first job has total job benefits of $45,600. The second job has total job benefits of $37,100.

PRACTICE PROBLEMS

3.16 Corina is offered a job with a salary of $48,940, and a benefits package totaling 25% of her salary. What are the total job benefits?

3.17 Xavier has a new job offer that includes the following benefits: pension $6,000, paid vacation: $5,600, health insurance $7,800. If the salary is $86,000, what are the total job benefits?

Net Job Benefits

You should also consider the expenses that you will incur with a job. Examples of job expenses are the cost to commute to the job, parking, uniforms or other clothing, equipment, and dues.

To find the *net job benefits*, subtract the job expenses from the total job benefits.

Net Job Benefits = Total Job Benefits − Job Expenses

Example: Nikki has been working at a new job for one year. She was hired with a salary of $26,450 and a 15% benefits package. Her job expenses for the first year were: commuting $1,500, professional dues $150, and continuing education $560. What were her net job benefits?

Solution: Find the total job benefits by multiplying the salary by the decimal form of the benefits package and adding the result to the salary. Add the expenses to find the total expenses and subtract from the total job benefits.

Benefit rate = 15% = 0.15

Employee benefits = $26,450 × 0.15 = $3,967.50

Total job benefits = $26,450 + $3,967.50 = $30,417.50

Job expenses = $1,500 + $150 + $560 = $2,210

Net job benefits = $30,417.50 − $2,210 = $28,207.50

Nikki's net job benefits are $28,207.50.

PRACTICE PROBLEMS

3.18 Melissa is offered a job with a salary of $38,395 and a benefits package of 15% of her salary. She estimates the following expenses: commuting $1,250, parking $240, professional dues $350. What are her estimated net job benefits?

3.19 Ron earned a salary of $74,800 last year. His employee benefits were 28% of his salary. His job expenses totaled $4,200. He has been offered a new job with a salary of $73,500 and benefits of 20% of his salary. He estimates his total job expenses to be $2,000. Will his current job or the new job offer the greatest net job benefits? How much more?

3.5 Take Home Pay

Many people are surprised at how much their gross pay is reduced by taxes and other deductions. You can evaluate the impact of deductions on your pay by finding the percentage of your gross pay that you actually take home.

To find the percentage of gross pay that you take home, divide net pay by gross pay and multiply by 100.

$$\text{Percentage of Gross Pay Taken Home} = \frac{\text{Net Pay}}{\text{Gross Pay}} \times 100$$

Example: Muhammed's gross pay is $860 per week. From his pay, the following deductions are taken: health insurance $25, federal income tax $108, and Social Security and Medicare taxes $48.59. What percentage of his gross pay does Muhammed take home? Round to the nearest percent.

Solution: Add all of the deductions and subtract the total from the gross pay to find the net pay. Divide the net pay by the gross pay and multiply by 100.

Total deductions = $25 + $108 + $48.59 = $181.59

Net pay = $860 − $181.59 = $678.41

$$\text{Percentage of Gross Pay Taken Home} = \frac{\$678.41}{\$860} \times 100 = 79\%$$

Muhammed takes home 79% of his gross pay.

PRACTICE PROBLEMS

3.20   Frances' gross pay is $1,980 per month. Her total deductions are $330. What percentage of her gross pay does Frances take home? Round to the nearest percent.

3.21   Iliana's monthly gross pay is $2,625. Federal withholding, Social Security, and Medicare taxes are deducted from her pay. If she is married and claims 2 withholding allowances, what percentage of her gross pay does Iliana take home? Use Figure 3.1 to find the federal withholding, and 5.65% for FICA tax. Round to the nearest percent.

Raises

Some employees are awarded a *raise* after they have worked for the company for a period of time or receive a good evaluation. Even a raise is impacted by deductions, and you will not take home the entire raise.

Example: Jillian earns $2,325 a month. She is married and claims three withholding allowances. Federal income tax, Social Security and Medicare taxes, and a $125 life insurance premium are deducted from her check. He employer gives her a 5% raise. How much do her gross pay and her net pay increase? Use Figure 3.1 to find the federal withholding, and 5.65% for FICA tax.

Solution: First, find Jillian's net pay by adding her deductions and subtracting them from her gross pay. Multiply her current gross pay by 5% to find the amount of her raise. Add the amount of the raise to her gross pay to find the new gross pay. Calculate her new deductions and find her net pay after the raise. Subtract her new net pay from her previous net pay to find how much her net pay increased.

Federal withholding (from Figure 3.1) = $28

Social Security and Medicare rate (FICA) = 5.65% = 0.0565

FICA tax = $2,325 × 0.0565 = $131.36

Total deductions = $28 + $131.36 + $125 = $284.36

Current net pay = $2,325 − $284.36 = $2,040.64

Raise = 5% = 0.05

Gross pay increase = $2,325 × 0.05 = $116.25

New gross pay = $2,325 + $116.25 = $2,441.25

Federal withholding (from Figure 3.1) = $40

FICA tax = $2,441.25 × 0.0565 = $137.93

Total deductions = $40 + $137.93 + $125 = $302.93

New net pay = $2,441.25 − $302.93 = $2,138.32

Net pay increase = $2,138.32 − $2,040.64 = $97.68

Jillian's gross pay increases $116.25, and her net pay increases $97.68.

PRACTICE PROBLEMS

3.22 Lacey is a married employee who claims 0 withholding allowances. She currently makes $2,400 per month. In addition to federal withholding tax, FICA tax and a state income tax of 2% are deducted from her pay. If she receives a 7% raise, how much will her gross pay and net pay increase? Use Figure 3.1 to find the federal withholding, and 5.65% for FICA tax.

3.23 Although Lacey received a 7% increase to her gross pay, what is the percent increase in her net pay?

Pre-Tax Deductions

Some employers offer employees the opportunity to take advantage of qualified *pre-tax deductions*. This benefit might be called a cafeteria plan, a salary reduction, or a Section 125 plan. Qualified pre-tax deductions reduce the amount of your taxable wages, which reduces the amount deducted for FICA tax and the amount withheld for federal income tax.

Example: From the preceding example, Jillian is planning to participate in her employer's cafeteria plan and have her life insurance deduction taken as a pre-tax deduction. How much will her take home pay increase per month and per year? Use Figure 3.1 to find the federal withholding, and 5.65% for FICA tax.

Solution: From the previous example, find the total she will pay in federal withholding tax and FICA tax on her gross pay after her raise. Find the taxable wages after the pre-tax deduction by subtracting the life insurance payment from Jillian's gross pay. Calculate federal withholding tax and FICA tax on the new taxable wages and add them. Subtract the total taxes with the pre-tax deduction from the original taxes to find the amount her take home pay will increase per month. Multiply the monthly increase in take home pay by 12 to find the yearly increase in take home pay.

Federal withholding on gross pay with raise
(from previous example) = $40

FICA tax on gross pay with raise (from previous example) = $137.93

Total tax = $40 + $137.93 = $177.93

Taxable income after pre-tax deduction = $2,441.25 − $125 = $2,316.25

Federal withholding = $24

FICA tax = $2,316.25 × 0.0565 = $130.87

Total tax on after pre-tax deductions = $24 + $130.87 = $154.87

Increase in monthly take home pay = $177.93 − $154.87 = $23.06

Increase in yearly take home pay = $23.06 × 12 = $276.72

Jillian's take home pay would increase $23.06 per month and $276.72 per year.

PRACTICE PROBLEM

3.24 Serenity earns $2,685 a month. She is married and claims 4
withholding allowances. Her company offers a cafeteria plan,
and Serenity has $250 worth of qualifying expenses that can be
pre-tax deductions. How much will Serenity's monthly and yearly
take home pay increase if she takes advantage of the pre-tax
deduction? Use Figure 3.1 to find the federal withholding and
5.65% for FICA tax.

Chapter

4

Banking

Individuals and businesses utilize the services of a bank or credit union to help manage their money. Banks offer savings and checking accounts where clients can deposit money safely and withdraw money to pay creditors. Banks may offer other savings and investment options as well as loans for when individuals or businesses need to borrow money.

4.1 Savings Accounts

Many people and businesses maintain a *savings account* at a bank or credit union. The individual or business deposits money with the bank, and the bank pays *interest* on the money in the account.

Simple Interest

The most basic form of interest is *simple interest*. It is interest that is paid on a fixed amount of money. To find simple interest, multiply the amount of money, or the *principal*, by the rate of interest and by the number of time periods.

$$\text{Interest} = \text{Principal} \times \text{Rate} \times \text{Time}$$
$$I = P \times R \times T$$

Example: How much simple interest is paid on $2,000 with a 2.5% annual interest rate over 3 years?

Solution: Substitute the values into the simple interest formula. Multiply the principal times the interest rate expressed as a decimal times the number of time periods.

Principal = $2,000, Interest rate = 2.5% yearly = 0.025 yearly, Time = 3 years

Interest = $2,000 × 0.025 × 3 = $150

The principal earns $150 in simple interest.

In practice, though, most savings accounts don't earn simple interest, because the bank adds the interest payment to your account and the new balance, including the interest, is used to calculate the next interest payment. Interest earned on the growing, or compounding, balance is called *compound interest* and will be discussed in the next section. An account that earned simple interest would continue to earn interest only on the original principal.

Banks may calculate interest semiannually, quarterly, monthly, or daily.

Semiannually: Two times per year, or every six months

Quarterly: Four times per year, or every three months

Monthly: 12 times per year

Daily: 360 or 365 times per year

The interest formula can be used to find how much a bank will pay in interest for a single interest period.

Example: Find the amount of interest paid for a $1,500 deposit for three months at 2% annual interest paid quarterly. What is the new account balance?

Solution: Substitute the values into the simple interest formula. Multiply the principal by the rate by the time. Since the interest rate is stated annually, and the time is three months, or one quarter of a year, the number of time periods is $\frac{1}{4}$. To find the new account balance, add the interest to the principal.

$$\text{Principal} = \$1,500, \text{ Rate of Interest} = 2\% = 0.02, \text{ Time period} = \frac{1}{4}$$

$$\text{Interest} = \$1,500 \times 0.02 \times \frac{1}{4} = \$7.50$$

Account balance = $1,500 + $7.50 = $1,507.50

The interest paid for three months is $7.50, and the new account balance is $1,507.50.

PRACTICE PROBLEMS

4.1 Tang deposited $650 in a savings account that pays 1.5% annual interest paid monthly. If he makes no further deposits or withdrawals from the account, how much interest will he earn in the first month?

4.2 What amount of interest is paid in the first quarter on $25,000 if the annual interest is 5% paid quarterly?

Compound Interest

Compound interest is earned when the interest due is added to the account balance, and then the new account balance is used for the principal for the next interest period. To calculate compound interest manually, calculate the simple interest for a single interest period, add it to the principal to find the new balance, and calculate the simple interest for the next period, using the new balance. Continue this process until all of the interest periods have been calculated.

Example: How much interest is earned after three years on $2,000 if a 2.5% annual interest rate is compounded annually?

Solution: Calculate the interest for the first year and add the interest to the principal. Calculate the interest for the second year and add the interest to the balance. Continue for the third year.

Principal = $2,000, Interest rate = 2.5% = 0.025, Time = 1 year

Interest for first year = $2,000 × 0.025 × 1 = $50

New account balance = $2,000 + $50 = $2,050

Interest for year 2 = $2,050 × 0.025 × 1 = $51.25

New account balance = $2,050 + $51.25 = $2,101.25

Interest for year 3 = $2,101.25 × 0.025 × 1 = $52.53

New account balance = $2,101.25 + $52.53 = $2,153.78

Total interest earned = $2,153.78 − $2,000 = $153.78

After three years, the $2,000 principal has earned $153.78.

In the first example with simple interest, in three years the $2,000 earned $150 in interest. With annual compounding, the $2,000 earned $153.78. While this may not seem like a lot of difference, many bank accounts are compounded daily. There will be a greater difference between simple and compound interest as the number of compounding periods per year increases.

PRACTICE PROBLEMS

4.3   If you deposit $1,500 in an account that pays 2.25% annual interest compounded semiannually, how much will you have in the account after one year?

4.4   Margo deposited $6,000 in a savings account that pays 3.5% annual interest compounded quarterly. How much interest will the account earn in the first year?

Manually calculating compound interest isn't a very efficient method, particularly as the number of compounding periods increase. A compound interest table can simplify the problem.

Compound Interest Table								
Rate Per Period								
Periods	0.25%	0.50%	0.75%	1%	1.25%	1.50%	2%	3%
4	1.010038	1.020151	1.030339	1.040604	1.050945	1.061364	1.082432	1.125509
6	1.015094	1.030378	1.045852	1.061520	1.077383	1.093443	1.126162	1.194052
7	1.017632	1.035529	1.053696	1.072135	1.090850	1.109845	1.148686	1.229874
8	1.020176	1.040707	1.061599	1.082857	1.104486	1.126493	1.171659	1.266770
20	1.051206	1.104896	1.161184	1.220190	1.282037	1.346855	1.485947	1.806111
30	1.077783	1.161400	1.251272	1.347849	1.451613	1.563080	1.811362	2.427262
40	1.105033	1.220794	1.348349	1.488864	1.643619	1.814018	2.208040	3.262038

Figure 4.1
Compound interest table.

Figure 4.1 shows the value of one dollar after it is compounded for various rates and time periods. Notice the increasing amount as the number of time periods and the interest rate increase. For example, a dollar that earns 0.25% interest per period will only grow to $1.11 in 40 interest periods, while a dollar that earns 3% interest per period will more than triple to $3.26 in the same time.

To calculate the interest, find the column that represents the interest rate per period and the row that represent the total number of compounding periods. Multiply the principal by the number that appears in the corresponding row and column.

To find the interest rate per period, divide the annual interest rate by the number of compounding periods per year.

$$\text{Interest Rate Per Period} = \frac{\text{Annual Interest Rate}}{\text{\# of Compounding Periods Per Year}}$$

Example: Find the account balance and the amount of interest after $2\frac{1}{2}$ years on $5,000 deposited into a savings account that pays 3% annual interest, compounded monthly.

Solution: Find the interest rate per period by dividing the interest rate by 12. Find the number of compounding periods by multiplying $2\frac{1}{2}$ by 12.

Use the table to find the multiplier. Multiply the principal by the multiplier from the table to find the account balance. Subtract the principal from the account balance to find the amount of interest.

Interest Rate Per Period $= \dfrac{3\%}{12} = 0.25\%$

Number of periods $= 2.5$ years $\times$ 12 periods per year $= 30$ periods

Multiplier from table $= 1.077783$

Account balance $= \$5,000 \times 1.077783 = \$5,388.92$

Interest earned $= \$5,388.92 - \$5,000 = \$388.92$

The account balance after $2\frac{1}{2}$ years is \$5,388.92, and \$388.92 was paid in interest.

PRACTICE PROBLEMS

4.5 How much money is in an account after 10 years if a single deposit of \$2,000 is made in an account that pays 4% annual interest, compounded quarterly?

4.6 Jane received a gift of \$3,500 when she was 13 years old. Her parents put the money in a savings account. The money was left untouched until Jane went to college five years later. The money earned 6% annual interest, compounded quarterly. How much money was in the account, and how much interest was earned in five years?

Another way to calculate compound interest is with the *compound interest formula*.

$$A = P(1 + r)^n$$

$A =$ account balance including interest

$P =$ original principal

$r =$ interest rate per period

$n =$ number of compounding periods

> **TIP**
>
> Remember to use the order of operations. Evaluate what is inside the parentheses first, then the exponent, and then multiply by *P*.

Example: Use the compound interest formula to find the account balance and the amount of interest after $2\frac{1}{2}$ years on $5,000 deposited into a savings account that pays 3% annual interest, compounded monthly.

Solution: Identify the interest rate per period and the number of compounding periods. To find the account balance, substitute the values into the compound interest formula and calculate. To find the amount of interest, subtract the principal from the new account balance.

$$\text{Interest Rate Per Period} = \frac{3\%}{12} = 0.25\% = 0.0025$$

$$\text{Number of compounding periods} = 2\frac{1}{2} \text{ years} \times 12 \text{ periods per year}$$

$$- 30 \text{ periods}$$

$A - P(1 + r)^n$

$A = \$5,000(1 + 0.0025)^{30} = \$5,000(1.0025)^{30}$

$A = \$5,388.92$

Interest earned $= \$5,388.92 - \$5,000 - \$388.92$

The account balance is $5,388.92 and the amount of interest earned is $388.92.

PRACTICE PROBLEMS

4.7 How much is in an account after five years that starts with $6,500 and earns 5% annual interest compounded semiannually?

4.8 How much interest is earned after 20 years on $15,000 earning 4% annual interest compounded quarterly?

Annual Percentage Yield

Interest rates can be very difficult to compare. One bank may pay interest at 1.5% compounded quarterly, while another may pay 1% compounded daily. The *annual percentage yield (APY)* or the *effective rate of interest* takes into account the effects of compounding. You would earn the same amount of interest in a year on two accounts with different interest rates and compounding periods if the APY for both accounts is the same.

One way to calculate the APY is to divide the interest earned in one year by the principal and change the decimal to a percent.

$$\text{APY} = \frac{\text{Interest Earned in One Year}}{\text{Principal}}$$

Example: Find the annual percentage yield to the nearest hundredth of a percent on an account with a $10,000 balance that pays 6% annual interest compounded quarterly.

Solution: Find the account balance after one year using Figure 4.1 or the compound interest formula. Find the interest earned by subtracting the principal from the account balance. Divide the interest earned by the principal to find the APY as a decimal. Convert to a percent by multiplying by 100.

Interest rate per period = 6% ÷ 4 = 1.5% = 0.015

Number of periods = 4

From Figure 4.1: multiplier = 1.061364

Account balance = $10,000 × 1.061364 = $10,613.64

Using the formula: $A = \$10,000(1 + 0.015)^4 = \$10,613.64$

Interest earned = $10,613.64 − $10,000 = $613.64

$$\text{APY} = \frac{\$613.64}{\$10,000} = 0.0614 = 6.14\%$$

The APY is approximately 6.14%

Another formula that converts the annual interest rate to the APY is

$$APY = (1 + r)^n - 1$$

r = interest rate per period

n = number of compounding periods

Example: Find the annual percentage yield to the nearest hundredth of a percent on an account that pays 6% annual interest compounded quarterly.

Solution: From the previous example, we know the interest rate per period is 0.015, and the number of periods is 4. Substitute the values into the formula and calculate the APY as a decimal. Convert to a percent.

$$APY = (1 + r)^n - 1$$
$$APY = (1 + 0.015)^4 - 1 = 0.0613635506 = 6.14\%$$

PRACTICE PROBLEMS

4.9 Joshua deposits $5,000 in an account that pays 3% annual interest compounded quarterly. What is the APY, to the nearest hundredth of a percent?

4.10 Find the annual percentage yield to the nearest hundredth of a percent on an account that pays an annual interest rate of 2% compounded monthly.

The annual percentage yield should be disclosed with any account information. When comparing two accounts, always compare the APY to see which account will pay more interest.

To find out how much interest you would earn on a savings account in a year, multiply the principal by the APY. This assumes that you do not deposit or withdraw any money during the year.

TIP

Remember that APY, or annual percentage yield, has already accounted for the effect of compounding.

Example: One bank offers a savings account that is compounded monthly with an APY of 4.5%. If you deposit $5,000 in the account, how much will you have after one year?

Solution: Multiply the principal by the APY expressed as a decimal.

APY = 4.5% = 0.045

Interest = $5,000 × 0.045 = $225

Account balance = $5,000 + $225 = $5,225

The account will have a balance of $5,225 after one year.

Practice Problems

4.11 Penny deposits $600 in an account with an APY of 1.5%. How much will she earn in interest in one year?

4.12 One bank offers a savings account with an APY of 4.5%, while another offers an account with an APY of 3.2%. How much more will you earn in one year on $1,000 in the account with the 4.5% APY?

4.2 Money Market and CD Accounts

Money market accounts and *CDs*, or *certificates of deposit*, are saving vehicles that many banks offer with a higher interest rate than a traditional savings account. The higher interest rate comes with some restrictions.

A money market account will typically require a much higher minimum balance, and may limit the number of withdrawals you can make in a given period of time.

A CD will have a minimum deposit, and you are required to leave the money on deposit for a specific amount of time. A *simple interest CD* allows for the interest on the CD to be paid out to the owner of the account with no penalty, although you will earn less interest on this type of CD because you lose the benefit of compound interest.

Example: How much interest will you earn on a one-year $5,000 simple interest CD with 4.5% interest paid quarterly? How much would you earn in interest each quarter?

Solution: You can use the simple interest formula to calculate one year of interest, and divide the yearly interest by 4 to find out how much you would earn each quarter. Or, you can use the simple interest formula to calculate interest for one quarter by using $\frac{1}{4}$ for the time, and then multiply the quarterly interest by 4 to find the yearly interest.

$I = P \times R \times T$

$P = \$5,000, R = 4.5\% = 0.045, T = 1$ year

$I = \$5,000 \times 0.045 \times 1 = \$225$

Quarterly interest $= \$225 \div 4 = \$56.25$

Using the second method:

$P = \$5,000, R = 4.5\% = 0.045$, and $T = \frac{1}{4}$

$I = \$5,000 \times 0.045 \times \frac{1}{4} = \$56.25$

Yearly interest $= \$56.25 \times 4 = \$225$

The simple interest CD would earn \$225 in one year, or \$56.25 per quarter.

PRACTICE PROBLEMS

4.13 Summer opens a five-year simple interest CD that earns 6% interest, paid semiannually. How much interest will she earn in five years if the principal is \$1,000?

4.14 Billy deposits \$25,000 in a simple interest CD that earns 5% interest, paid quarterly. How much interest will Billy earn each quarter?

If you withdraw money from a CD before the end of the term, or the *maturity date*, you will pay a penalty. The penalty for early withdrawal is usually expressed as a loss of a certain number of month's interest.

Example: Dan invested \$10,000 in a five-year CD that paid 4.1% interest. The CD has an early withdrawal penalty of six months' simple interest. After two years, he cashed out the CD. How much does he pay for the penalty?

Solution: Use the interest formula to calculate six months, or half a year of interest.

$$I = P \times R \times T$$

$$P = \$10,000, R = 4.1\% = 0.041, T = \frac{1}{2} \text{ year}$$

$$I = \$10,000 \times 0.041 \times \frac{1}{2} = \$205$$

Dan will pay $205 for the early withdrawal penalty.

PRACTICE PROBLEMS

4.15 Jack has $4,000 in a two-year simple interest CD earning 5% annual interest. The bank charges an early withdrawal penalty of three months simple interest. If Jack cashes out the CD early, how much is the early withdrawal penalty?

4.16 Victoria has a $2,000 simple interest CD that earns 3.5% annual interest. She cashes out the CD after one year and pays a penalty of two months interest. How much does she receive from the bank when she cashes out the CD?

4.3 Annuities

Most people do not place a lump sum of money in a savings account and leave it there. The goal for a saving account is usually to continue to add money to the account. An *annuity* is the process of making or receiving a series of fixed payments, such as saving an amount of money by making regular deposits into an account, or the process liquidating an account by making regular withdrawals. Specific examples of annuities include paying rent, making a car payment, or saving a fixed amount at certain intervals to be withdrawn later for retirement or a child's education.

The insurance industry also offers a product that is called an annuity. It is based on the mathematics of future value and present value that will be discussed. However, an insurance company annuity is a contract whereby you make a series of regular deposits over time, and the insurance company guarantees you a series of fixed payments beginning at a certain point and lasting for the rest of your life.

Future Value of an Ordinary Annuity

The *future value of an ordinary annuity* is the money that will be in an account after a series of regular deposits, including the interest that the money has made.

One way to find the future value of an ordinary annuity is to use a table like the one shown in Figure 4.2. To use the table, find the multiplier that represents the rate of interest per period and the number of periods. Multiply that number by the amount of the regular deposits.

FUTURE VALUE OF ORDINARY ANNUITY									
Rate Per Period									
Periods	0.25%	0.50%	0.75%	1.00%	1.50%	2.00%	2.50%	3.00%	4.00%	5.00%
4	4.01503	4.03010	4.04523	4.06040	4.09090	4.12161	4.15252	4.18363	4.24646	4.31013
10	10.11325	10.22803	10.34434	10.46221	10.70272	10.94972	11.20338	11.46388	12.00611	12.57789
11	11.13854	11.27917	11.42192	11.56683	11.86326	12.16872	12.48347	12.80780	13.48635	14.20679
12	12.16638	12.33556	12.50759	12.68250	13.04121	13.41209	13.79555	14.19203	15.02581	15.91713
16	16.30353	16.61423	16.93228	17.25786	17.93237	18.63929	19.38022	20.15688	21.82453	23.65749
20	20.48220	20.97912	21.49122	22.01900	23.12367	24.29737	25.54466	26.87037	29.77808	33.06595
24	24.70282	25.43196	26.18847	26.97346	28.63352	30.42186	32.34904	34.42647	39.08260	44.50200
40	42.01320	44.15885	46.44648	48.88637	54.26789	60.40198	67.40255	75.40126	38.02552	120.7998

Figure 4.2
Future value of an ordinary annuity table.

Example: To save for their child's college education, Abel and Teresa deposit $400 per quarter in an account, beginning when their child is eight years old. The account earns 4% interest, compounded quarterly. How much will be in the account when the child turns 18?

Solution: Find the interest rate per period by dividing the interest rate by the number of compounding periods per year. Find the number of periods by multiplying the number of years by four quarters per year. Identify the multiplier from the table the represents the interest rate per period and the number of periods. Multiply the multiplier by the amount deposited each month.

Interest rate per period $= \dfrac{4\%}{4} = 1\%$

Number of years $= 18 - 8 = 10$ years

Number of periods $= 10 \times 4 = 40$

Multiplier $= 48.88637$

Amount in account after 10 years $= \$400 \times 48.88637 = \$19{,}554.55$

There will be $19,554.55 in the account after 10 years.

Practice Problems

4.17 Andre saves $200 per month in an account that pays 3% annual interest, compounded monthly. How much will be in the account after two years?

4.18 Sylvia saves $1,000 quarterly and puts it into an account that earns 8% annual interest, compounded quarterly. How much will be in the account after four years?

Present Value of an Ordinary Annuity

The *present value of an ordinary annuity* is the money that is required to be in an account in order to make a series of payments from the account. Interest is still being earned on the money that is left in the account while the payments are being made, and that interest can be used for future payments.

One way to find the present value of an ordinary annuity is to use a table like the one in Figure 4.3. To use the table, find the multiplier that represents the rate of interest per period and the number of periods. Multiply that number by the amount of the regular withdrawals.

Example: Abel and Teresa want to have enough money in their child's college account to make quarterly payments of $2,000 for their child's education, for four years. How much money do they need in the college account by the time their child starts college if their account earns 4% interest, compounded quarterly?

	PRESENT VALUE OF ORDINARY ANNUITY									
	Rate Per Period									
Periods	0.25%	0.50%	0.75%	1.00%	1.50%	2.00%	2.50%	3.00%	4.00%	5.00%
4	3.97512	3.95050	3.92611	3.90197	3.85438	3.80773	3.76197	3.71710	3.62990	3.54595
10	9.86386	9.73041	9.59958	9.47130	9.22218	8.98259	8.75206	8.53020	8.11090	7.72173
11	10.83677	10.67703	10.52067	10.36763	10.07112	9.78685	9.51421	9.25262	8.76048	8.30641
12	11.80725	11.61893	11.43491	11.25508	10.90751	10.57534	10.25776	9.95400	9.38507	8.86325
16	15.66504	15.33993	15.02431	14.71787	14.13126	13.57771	13.05500	12.56110	11.65230	10.83777
20	19.48449	18.98742	18.50802	18.04555	17.16864	16.35143	15.58916	14.87747	13.59033	12.46221
24	23.26598	22.59287	21.88915	21.24339	20.03041	18.91393	17.88499	16.93554	15.24696	13.79864
40	38.01986	36.17223	34.44694	32.83469	29.91585	27.35548	25.10278	23.11477	19.79277	17.15909

Figure 4.3
Present value of an ordinary annuity table.

Solution: Find the interest rate per period by dividing the interest rate by the number of compounding periods per year. Find the number of periods by multiplying the number of years by four quarters per year. Identify the multiplier from the table that represents the interest rate per period and the number of periods. Multiply the multiplier by the amount withdrawn each quarter.

$$\text{Interest rate per period} = \frac{4\%}{4} = 1\%$$

Number of periods = 4 years $\times$ 4 periods per year = 16 periods

Multiplier = 14.71787

Amount needed = $2,000 $\times$ 14.71787 = $29,435.74

Abel and Teresa need to have $29,435.74 in the account by the time the child starts college.

In the example in the previous section, Abel and Teresa were saving $400 per quarter for 10 years, earning 4% annual interest. Using the table in Figure 4.2, we calculated that they would have $19,554.55 in the account after 10 years. However, the current example shows that to receive $2,000 quarterly from the account for four years, they need $29,435.74 in the account. These two examples illustrate that to meet their goal Abel and Teresa need to save more money each quarter, earn a higher rate of interest, or they should have started saving earlier.

Practice Problems

4.19 To supplement his retirement income, Charles wants to receive $5,000 semiannually for 10 years. If the money will be invested where it earns 8% annual interest compounded semiannually, how much money needs to be in the account by the time he begins to withdraw money?

4.20 If your money earns 2% compounded quarterly, how much money do you need in an account to receive quarterly payments of $200 for six years?

4.4 Checking Accounts

Individuals and businesses usually have a *checking account* that they deposit money into, and then write checks to make payments. Often a checking account will have an ATM or debit card associated with it to withdraw or deposit money at an ATM machine or to make purchases without writing a check.

Most banks now offer online banking where customers can view statements and transactions online, transfer money between savings and checking accounts, and pay bills.

Deposits

To deposit money in a checking account, you must fill out a *deposit slip* showing how much money in cash and checks that you are depositing. You can also use a line on the deposit slip to direct to the bank if you want to receive cash back, which is deducted from the subtotal to find the total amount deposited.

A completed deposit slip for a deposit of $14.86 in cash and checks of $46.78 and $168.46 is shown in Figure 4.4. Note that no cash back was received, so the total amount was deposited.

```
Jan Summers                                      ☐ CASH ▶ [        14].[8][6]
1805 Popular Ln.                                   37-885119-3642
Sunnyberg, PA                                      1737449460
                                                          ___▶ [        46].[7][8]
                                                          ___▶ [       168].[4][6]
DATE __April 5, 20--_____                            ___▶ [          ].[ ][ ]
DEPOSITS MAY NOT BE AVAILABLE FOR IMMEDIATE WITHDRAWAL  SUB TOTAL ▶ [ 230].[1][0]
                                                        *LESS CASH ▶ [   0].[0][0]
                                                        RECEIVED
_____                          $ [       230].[1][0]
       *SIGN HERE FOR CASH RECEIVED (IF REQUIRED)

A Anchor Bank
⑈580800393⑈:0000007449460⑈
```

Figure 4.4
Sample deposit slip.

On the reverse side of most deposit slips is space where you can fill in additional checks if you have more than will fit on the front side. If you use the back, find the total of the checks on the back and enter it on the space marked Total From Reverse.

Example: Marion deposits the following in her checking account: 6 twenties, 3 quarters, 2 dimes, and 2 checks, $38.67 and $168.46. She requests cash back of 10 five-dollar bills. What is the total of her deposit?

Solution: Find the total cash she is depositing and add to it the amount of the checks. Find the amount of money she wants in cash back and subtract it from the deposit.

$6 \times \$20 = \$120$

$3 \times \$0.25 = \$0.75$

$2 \times \$0.10 = \$0.20$

Total cash $= \$120 + \$0.75 + \$0.20 = \$120.95$

Subtotal $= \$120.95 + \$38.67 + \$168.46 = \$328.08$

Cash back $= 10 \times \$5 = \$50$

Total deposit $= \$328.08 - \$50 = \$278.08$

Marion's deposit is $278.08.

PRACTICE PROBLEMS

4.21 Jackson deposits three checks in the amounts of $468.22, $75.43, and $122.89. He requests $150 cash back. What is the amount of his deposit?

4.22 Jim deposits five $50-dollar bills, ten $20-dollar bills, six $5-dollar bills, and $2.34 in change. What is the amount of his deposit?

Check Register

To keep track of money that is deposited and withdrawn from a checking account, you should record transactions in the *check register*. The check register keeps a running balance of your account.

A sample checkbook register is shown in Figure 4.5. The beginning balance on May 1 was $385.23. A deposit of $500 was made on May 2. Check 387 was written on May 3 for $22.14, and on online payment of $129.34 was made on May 4. An ATM withdrawal of $150 was made on May 6, and the ending balance on May 6 was $583.75.

CHECK REGISTER

NUMBER	DATE	DESCRIPTION OF TRANSACTION	PAYMENT/DEBIT (–)	✔ T	FEE (–)	DEPOSIT/CREDIT (+)	BALANCE	
							$385	23
DEP	5/2	Deposit				$500 00	885	23
387	5/3	Hank's Hardware Store	22 14				863	09
ONLINE	5/4	Credit Bank	129 34				733	75
ATM	5/6	Cash	150 00				583	75

Figure 4.5
Sample check register.

Example: On September 20, Cindy's checking account balance was $2,855.43. She wrote check 1821 on September 21 to Cycorp Electric for $637.46. She directed her bank's online bill pay program to make a payment of $750 to Property Management, Inc. for rent on September 22. She used her debit card on September 23 at Milli's Restaurant for $22.81. She made a deposit of $500 on September 24. Update the check register and find her balance at the end of September 24th.

Solution: To keep a running balance, to the previous balance add deposits and subtract checks, payments, fees, and other debits.

Balance on 9/21 = $2855.43 − $637.46 = $2,217.97

Balance on 9/22 = $2,217.97 − $750 = $1,467.97

Balance on 9/23 = $1,467.97 − $22.81 = $1,445.16

Balance on 9/24 = $1,445.16 + $500 = $1,945.16

Cindy's final balance on September 24th is $1,945.16.

If you keep the register updated with deposits, checks, ATM transactions, debit card transactions, and any online banking activity, you will know how much you have in your checking account at any given time. If you don't keep an accurate balance, you might spend money that you do not have in your account, and your account will be *overdrawn*. Banks will often charge an overdraft fee for accounts that are overdrawn. In addition, if the bank chooses not to honor your check, this can hurt your standing with the person or business to which you wrote the check. Some banks offer checking accounts with *overdraft protection*, which means the bank will honor your check (even if there is not enough money in your checking account) by extending a loan or withdrawing the money from your savings account.

PRACTICE PROBLEMS

4.23 Ginger had a beginning checking account balance of $122.31. She made a deposit of $653.82, and made three debit purchases of $25.69, $99.86, and $43.82. What was Ginger's final balance in the check register?

4.24 George had a check register balance of $1,486.22. He deposited $435 one day, and made an ATM withdrawal of $200 the next. He made an online payment of $305, and a debit card purchase of $75.22. What was George's final balance?

Reconciliation

Each month you will receive a statement from the bank by mail or to view online showing the transactions on your checking account for the previous month.

The account balance on the bank statement will probably not be the same as the account balance in your check register. Your check register probably does not show any interest you earned or fees charged. In addition, there may be checks that you have recorded in your check register that have not cleared the bank, called *outstanding checks*, and there may be deposits you have made that were deposited after the period of time covered on the statement, called *outstanding deposits*. The process of verifying transactions and adjusting the check register and the bank's balance to account for these differences is called *reconciling* the bank statement, or balancing your checkbook.

To reconcile your bank statement, follow these steps:

1. Check off transactions that appear on both the bank statement and the check register. In your checkbook register there should be a column provided to place a check in for transactions that have been posted. Verify that the transactions are for the exact same amount. Make any corrections to the check register, including updating the running balance.

2. Update your check register with any transactions that appear on the bank statement that you did not record. Once recorded, check them off on the bank statement and in the check register.

3. Locate any deposits that appear in your check register, but not on the statement. Add them up to find the *outstanding credit transactions*.

4. Locate any checks or other debits such as ATM withdrawals or debit card transactions that you recorded but are not on the bank statement. Add them up to find the *outstanding debit transactions*.

5. Use the following formula to find the adjusted bank balance:

   Adjusted Bank Balance = Bank Statement Balance +
   Outstanding Credit Transactions − Outstanding Debit Transaction

The adjusted bank balance should agree with the new balance in your check register. If it does not, check your calculations in the reconciliation steps as well as the running balance in your check register.

Example: Leah's bank statement shows a balance of $4,204.87. There are three outstanding checks for $22.49, $483.81, and $50. There is one outstanding deposit of $250. Reconcile Leah's bank statement.

Solution: Add the outstanding checks to find the outstanding debit transactions. From the bank statement balance, add the outstanding credit (deposit) and subtract the total outstanding debits (checks).

Outstanding debit transactions = $22.49 + $483.81 + $50 = $556.30

Adjusted bank balance = $4,204.87 + $250 − $556.30 = $3,898.57

The adjusted bank balance is $3,898.57.

PRACTICE PROBLEMS

4.25 Jun receives his May bank statement, and the bank's balance is $231.85. He notices the following outstanding transactions: checks of $11.43 and $121.86 and deposits of $589.22 and $263. What is the adjusted bank balance?

4.26 Your monthly bank statement shows a balance of $984.39. You found the following outstanding checks: $45.83, $75, and $82.19. The outstanding deposits were $525 and $84.39. What is the adjusted bank balance?

Chapter

5

Credit Cards

When you use a credit card, you are borrowing money. The credit card company pays the merchant, and then you pay back the credit card company. Depending on how quickly you pay back the credit card company and the specific agreement you have with the company, you may pay a sizable amount of interest.

5.1 Credit Card Disclosures

Credit card companies are required to disclose the terms and conditions of the credit card account, and they are required to place some specifics in a table so that you can understand the terms and compare different credit card accounts (see Figure 5.1).

Required Disclosure	Example
Annual Percentage Rate for Purchases	**8.99%, 10.99% or 13.99%** introductory APR for 6 months based on credit worthiness. After that the APR will be **15.99%**. This rate will vary with the Prime Rate.
APR for Balance Transfers	**0%** for the first 6 months. After that the APR will be **16.99%**. This rate will vary with the Prime Rate.
APR for Cash Advances	**22.99%** This rate will vary with the Prime Rate.
Penalty APR and When it Applies	**28.99%** This APR may be applied to your account if you: 1) Make a late payment; 2) Go over your credit limit; 3) Make a payment that is returned; or 4) Do any of the above on another account you have with us. **How Long Will the Penalty APR Apply?:** The penalty APR will apply until you make 4 consecutive minimum payments by the due date.
How to Avoid Paying Interest on Purchases	Your due date is at least 25 days after the close of each billing cycle. No interest is due if you pay your entire balance by the due date.
Minimum Interest Charge	Interest charges will be no less than $1.50.
Set-up and Maintenance Fee	
• **Annual Fee**	$40
• **Account Set-Up Fee**	$25 (one-time fee)
• **Additional Card Fee**	$10 annually (if applicable)
Transaction Fees	
• **Balance Transfer**	$5 or 3% of the amount transferred, whichever is greater
• **Cash Advance**	$5 or 3% of the amount transferred, whichever is greater
• **Foreign Transaction**	2% of each transaction
Penalty Fees	
• **Late Payment**	$29
• **Over the Credit Limit**	$29
• **Returned Payment**	$30

How we will calculate your balance: We will use the average daily balance method (including new purchases).
Loss of Introductory APR: We may end your introductory APR and apply the penalty APR if you are more than 30 days late with your payment.

Figure 5.1
Sample credit card disclosure table.

The first four rows of the table give information about the *annual percentage rate*, or APR, that is used to calculate interest. Typically, interest rates for balance transfers and cash advances will be higher than those for purchases. Some credit cards come with variable rates—rates that can change as certain economic indexes change.

The credit card company is required to tell you how to avoid paying interest charges. A *grace period* is the time in which you can pay your bill in full without incurring any finance charges. With most credit cards, there is no grace period for cash advances and balance transfers.

The remainder of the table details various fees. If you plan to pay off the balance on your credit card each month, then look for cards with fewer fees or no annual fee. If you plan to carry a balance on your credit card, a card with an annual fee will generally offer a lower interest rate.

Below the table will be a statement of how your balance is calculated. There are three common methods used that will be discussed in the next section.

Example: Jim is planning to open a new credit card account. The card has the terms and conditions outlined in Figure 5.1. He plans to transfer a $1,225 balance from his old credit card. In addition, in the first month, he will have to pay the annual fee, the account set-up fee, and the balance transfer fee. If he makes no purchases in the first month, what is his first month's credit card balance?

Solution: Multiply the amount of the balance transfer by the balance transfer fee. Add the amount of the balance transfer and all of the fees for the account.

Balance transfer fee = $5 or 3% = 0.03

Balance transfer fee = $1,225 × 0.03 = $36.75

Annual fee = $40

Account set-up fee = $25

Total fees = $36.75 + $40 + $25 = $101.75

Credit card balance = $1,225 + $101.75 = $1,326.75

Jim's first month credit card balance is $1,326.75.

Getting started with a new credit card can be costly, but interest that you pay on the balance of the credit card is where the costs can become significant.

Example: Suzanne opened a new credit card account. She paid an annual fee of $50 and a 4% balance transfer fee on $580. Over the next year, she paid an average of $45.83 per month in interest charges. What was the total cost of credit in the first year?

Solution: Calculate the balance transfer fee by multiplying the amount of balance transfer by the fee percent. Multiply the average monthly interest charge by 12 to find the yearly cost of interest. Add the annual fee, balance transfer fee, and the total interest.

Balance transfer fee percentage = 4% = 0.04

Balance transfer fee = $580 × 0.04 = $23.20

Annual interest = $45.83 × 12 = $549.96

Total cost of credit = $23.20 + $549.96 + $50 = $623.16.

The total cost of credit for one year was $623.16.

PRACTICE PROBLEMS

5.1   If Moria opens a credit card account with an additional card with the terms outlined in Figure 5.1, and then transfers a balance of $2,480, what will be the balance on her first bill?

5.2   Janeine has a credit card with an annual fee of $75. She paid a balance transfer fee of 5% on $3,400. She paid an average of $68.43 per month in interest for the first year. What is the total cost of credit for the first year?

5.2 Finance Charges

If you don't pay your entire credit card balance by the due date, you will be charged interest, called a *finance charge*. In addition, you will lose your grace period and all purchases will be subject to finance charges from the day of purchase until the credit card balance is paid in full.

The credit card disclosure statement gives the annual percentage rate (APR), but the finance charges are calculated using a monthly or daily *periodic rate*. To find the daily and monthly periodic rates, divide the APR by 12 or 365 and round to the nearest ten-thousandth.

- Monthly Periodic Rate $= \dfrac{\text{APR}}{12}$

- Daily Periodic Rate $= \dfrac{\text{APR}}{365}$

To find the periodic finance charge, multiply the balance subject to a finance charge by the periodic rate and by the number of periods.

- Periodic Finance Charge = Balance Subject to Finance Charge × Periodic Rate × Number of Periods

Example: Badal has a credit card with a balance of $483.22 subject to a finance charge. If the APR is 16%, what is the finance charge with a monthly periodic rate? What is the finance charge with a daily periodic rate if there are 31 days in the billing cycle?

Solution: Divide the APR by 12 and by 365 to find the monthly and daily periodic rates. To find the finance charge for the monthly periodic rate, multiply the balance by the monthly periodic rate and one period. To find the finance charge for the daily periodic rate, multiply the balance by the daily periodic rate and by 31, the number of days in the billing cycle.

Monthly periodic rate = 16% ÷ 12 = 1.3333% = 0.013333

Periodic finance charge = $483.22 × 0.013333 × 1 = $6.44

Daily period rate = 16% ÷ 365 = 0.0438% = 0.000438

Periodic finance charge = $483.22 × 0.000438 × 31 = $6.56

The finance charge with a monthly periodic rate is $6.44. The finance charge with a daily periodic rate is $6.56.

PRACTICE PROBLEMS

5.3 Max's credit card has a balance of $4,329.18 subject to a finance charge. His credit card has an APR of 18% and uses a monthly periodic rate. What is his current month's finance charge?

5.4 Juan must pay a finance charge on a balance of $822.49. If his credit card uses a daily periodic rate on an APR of 15%, what is his finance charge for a 30-day billing cycle?

While all credit card companies calculate the periodic finance charge using the same formula, there are a variety of methods by which credit card companies calculate the balance that is subject to a finance charge.

Previous Balance Method

The *previous balance method* uses the balance on the last billing date of the previous month as the balance subject to a finance charge. Any activity on the account during the current month is not included for calculating the finance charge.

Use the Periodic Finance Charge formula with the previous balance to find the finance charge. The new balance on the credit card will be the previous balance plus the sum of the finance charge, new purchases, and fees minus the sum of the payments and credits.

• New Balance = Previous Balance + (Finance Charge +
  New Purchases + Fees) − (Payments + Credits)

Example: Jasmine's previous balance on her credit card was $394.75. During the billing cycle, she made $145.82 worth of new purchases and had credits totaling $283.44. Her credit card company uses the previous balance method with a daily periodic rate and an APR of 15%. If there are 30 days in the billing cycle, calculate her finance charge and her new balance.

Solution: Divide the APR by 365 to find the daily periodic rate. Multiply the daily periodic rate by the previous balance and 30 days in the billing cycle to find the finance charge. Add the previous balance to the sum of the finance charge and new purchases and subtract the credits to find the new balance.

Daily periodic rate = 15% ÷ 365 = 0.0411% = 0.000411

Finance charge = $394.75 × 0.000411 × 30 = $4.87

New balance = $394.75 + ($4.87 + $145.82) − $283.44 = $262

Jasmine's finance charge is $4.87, and her new balance is $262.

PRACTICE PROBLEMS

5.5 Simeon's credit card has a previous balance of $4,821.13. During the current billing cycle, he made purchases totaling $529.74 and had credits totaling $1,485. What is Simeon's finance charge and new balance if the credit card has an APR of 20%, uses the previous balance method and a monthly periodic rate?

5.6 Johann's credit card uses the previous balance method using a daily periodic rate and an APR of 12%. For the 30-day billing cycle, Johann had a previous balance of $692.28, new purchases of $238.92, and credits of $70. What are Johann's finance charge and the new balance on the credit card?

Adjusted Balance Method

The *adjusted balance method* uses the previous balance minus the payments and credits applied during the current billing period. Purchases made and fees applied during the current billing period are not included in the adjusted balance.

- Adjusted Balance = Previous Balance − (Payments + Credits)
- New Balance = Adjusted Balance + Finance Charge + New Purchases + Fees

Example: Jasmine's previous balance on her credit card was $394.75. During the billing cycle, she made $145.82 worth of new purchases and had credits totaling $283.44. Her credit card company uses the adjusted balance method with a daily periodic rate and an APR of 15%. If there are 30 days in the billing cycle, calculate her finance charge and her new balance.

Solution: Divide the APR by 365 to find the daily periodic rate. Subtract the credits from the previous balance to find the adjusted balance. Multiply the daily periodic rate by the adjusted balance and 30 days in the billing cycle to find the finance charge. Add the adjusted balance to the finance charge and new purchases and fees to find the new balance.

Daily periodic rate = 15% ÷ 365 = 0.0411% = 0.000411

Adjusted balance = $394.75 − $283.44 = $111.31

Finance charge = $111.31 × 0.000411 × 30 = $1.37

New balance = $111.31 + $1.37 + $145.82 = $258.50

Jasmine's finance charge is $1.37, and her new balance is $258.50.

Notice that if there are payments or credits on the account, the adjusted balance method will lead to lower finance charges than the previous balance method.

PRACTICE PROBLEMS

5.7   What are Simeon's finance charge and new balance if the credit card company uses the adjusted balance method? See practice problem 5.5.

5.8   What are Johann's finance charge and new balance if the credit card company uses the adjusted balance method? See practice problem 5.6.

Average Daily Balance Method

The *average daily balance* method is the most common method used to calculate the balance subject to finance charges. In this method, the average daily balance is found by adding the balances at the end of each day and dividing by the number of days in the billing period. The periodic rate is applied to the average daily balance.

There are two forms of average daily balance: one that includes new purchases, called the *average daily balance including new purchases* method and one that excludes new purchases from the daily balance, called the *average daily balance excluding new purchases* method.

Use the following formulas to calculate finance charges using the average daily balance including new purchases.

- Daily Balance with New Purchases = Beginning Balance − (Payments + Credits) + (Purchases and Fees)

- Average Daily Balance = $\dfrac{\text{Sum of Daily Balances}}{\text{Number of Days in the Billing Cycle}}$

- New Balance = Previous Balance + (Finance Charges + New Purchases + Fees) − (Payment + Credits)

Example: Jonisha's credit card statement showed the following activity on her credit card: 7/1 previous balance $623.15, 7/4 purchase $48.23, 7/11 purchase $18.42, 7/20 payment $50, 7/26 purchase $29.43. Jonisha's credit card company uses the average daily balance method including new purchases and an APR of 17% using a monthly periodic rate. Find the finance charge and the new balance.

Solution: Organize the transactions and daily balances in a table, like the one shown in Table 5.1. Add the daily balances and divide by 31 to find the average daily balance. Multiply the average daily balance by the monthly periodic rate to find the finance charge. To find the new balance, find the sum of the previous balance and the finance charge, new purchases, and fees and subtract any payments or credits.

Table 5.1 Daily Balances Including New Purchases

Date	Transaction	Balance	# of days	Sum of Daily Balances
7/1		$623.15	3	$ 1,869.45
7/4	+ $48.23	$671.38	7	$ 4,699.66
7/11	+ $18.42	$689.80	9	$ 6,208.20
7/20	− $50	$639.80	6	$ 3,838.80
7/26	+ $29.43	$669.23	6	$ 4,015.38
Total			31	$20,631.49

Average daily balance $= \dfrac{\$20{,}631.49}{31} = \$665.53$

Monthly periodic rate $= 17\% \div 12 = 1.4167\% = 0.014167$

Finance charge $= \$665.53 \times 0.014167 = \$9.43$

New balance $= \$623.15 + (\$9.43 + \$48.23 + \$18.42 + \$29.43) - \$50 = \$678.66$

Jonisha's finance charge is $9.43, and her new balance is $678.66.

If the credit card company uses the average daily balance excluding new purchases the same method is used, but without the addition of new purchases.

Use the following formula to calculate the daily balance excluding new purchases. The average daily balance and the new balance are calculated the same as the average daily balance including new purchases.

Daily Balance Excluding New Purchases = Beginning Balance −
(Payments + Credits) + Fees

Example: What would Jonisha's finance charge and new balance be if the average daily balance excluding new purchases method is used? See the previous example.

Solution: Organize the transactions and daily balances excluding new purchases in a table, like the one shown in Table 5.2. Add the daily balances and divide by 31 to find the average daily balance. Multiply the average daily balance by the monthly periodic rate to find the finance charge. To find the new balance, find the sum of the previous balance and the finance charge, new purchases and fees and subtract any payments or credits.

Table 5.2 Daily Balances Excluding New Purchases

Date	Transaction	Balance	# of days	Sum of Daily Balances
7/1		$623.15	19	$11,839.85
7/20	− $50	$573.15	12	$ 6,877.80
Total			31	$18,717.65

$$\text{Average daily balance} = \frac{\$18,717.65}{31} = \$603.80$$

Monthly periodic rate $= 17\% \div 12 = 1.4167\% = 0.014167$

Finance charge $= \$603.80 \times 0.014167 = \$8.55$

New balance $= \$623.15 + (\$8.55 + \$48.23 + \$18.42 + \$29.43) - \$50 = \$677.78$

Jonisha's finance charge is $8.55, and her new balance is $677.78.

Notice that if there are purchases made during the billing cycle, the average daily balance excluding new purchases method will lead to lower finance charges than the average daily balance including new purchases method.

PRACTICE PROBLEMS

5.9 Jane's credit card statement showed the following transactions: 4/1 beginning balance $1,543.22, 4/3 purchase $127.89, 4/5 credit $28.13, 4/8 fee $40, 4/19 purchase $22.83, 4/22 payment $150. If her credit card charges an APR of 18% using a monthly periodic rate and the average balance including new purchases, what is Jane's finance charge and new balance?

5.10 From the previous problem, if Jane's credit card company uses the average daily balance excluding new purchases, what is Jane's finance charge and new balance?

5.3 Cash Advances

Credit card companies have made it easy to use your credit card to get cash, called a *cash advance*. Most credit cards can be used to draw money out at an ATM, and many companies will mail you checks that you can use to get cash.

Cash advances come with a high price. Typically, there is no grace period for cash advances, so finance charges, usually at a very high interest rate, accumulate from the day you receive the cash. In addition, most credit cards charge an additional fee for the cash advance.

To find the finance charges on a cash advance, use the same formula for the periodic finance charge. In most cases, a daily periodic rate is applied to the number of days in the billing cycle since the advance was made. The total finance charge will also include any fee charged for the cash advance.

- Periodic Finance Charge = Balance Subject to Finance Charge × Periodic Rate × Number of Periods
- Total Finance Charge = Periodic Finance Charge + Fees

Example: On June 5, Paul took a $500 cash advance on his credit card. The credit card company charges a 5% cash advance fee plus an APR of 23% using a daily periodic rate from the day the advance was made. What is the total finance charge for the month of June for the cash advance?

Solution: Find the cash advance fee by multiplying the cash advance amount by 5%. Calculate the daily periodic rate by dividing the APR by 365. Multiply the cash advance by the daily periodic rate and the number of days since the cash advance was taken to find the finance charge. Add the fee and the finance charge to find the total finance charge.

Cash advance fee percentage = 5% = 0.05

Cash advance fee = $500 × 0.05 = $25

Daily periodic rate = 23% ÷ 365 = 0.0630% = 0.000630

Number of days = 26 days

Finance charge = $500 × 0.000630 × 26 = $8.19

Total finance charge = $25 + $8.19 = $33.19

The total finance charge for the cash advance is $33.19.

PRACTICE PROBLEMS

5.11   Julius used his credit card in an ATM machine to get a $300 cash advance. His credit card company charges a $10 cash advance fee and a daily periodic interest rate on an APR of 22%. If Julius had the cash advance for 26 days, what was the total finance charge for the cash advance?

5.12   Aakar took a cash advance of $400 on his credit card. The company
charges a 3% fee and a daily periodic rate on an APR of 21%. If
Aakar had the cash advance for 40 days, how much does he pay back
in total to pay off the cash advance and the total finance charge?

If you have made purchases and received a cash advance on your credit
card, your finance charges are calculated separately, since there will
likely be different interest rates for purchases and for cash advances.

Historically, credit card companies applied any payments to the balance
with the lower interest rate first, so that if you didn't pay off your entire
balance, you continued to carry the balance subject to the higher interest
rate. The Credit Card Accountability Responsibility and Disclosure Act,
which went into effect in 2010, ended this practice, and provided other
consumer protections for credit card holders.

5.4 Debt Management

Many people get into huge financial problems due to credit card debt.
Credit cards are easy to use and as a result, many people overspend.
Furthermore, credit card companies only require a small *minimum
payment* each month, some as little as 1% plus the current finance
charges.

Minimum Credit Card Payments

Making only the minimum payment and continuing to use the credit card
will usually create an increasing credit card balance over time.

Example: Nicolai has a credit card with an 18% APR. Monthly periodic
finance charges are calculated using the previous balance method. His
balance on January 1 is $1,500, and he charges an average of $150 per
month on his credit card. He makes the minimum payment on his card
which is 2% of his current balance, rounded to the nearest dollar. What
is the balance on his card after six months? How much has he made in
payments in six months?

Solution: Make a table to show the finance charges, new purchases, payments and balance for each month (see Figure 5.2).

Nicolai's balance at the end of six months is $2,324.11. He has made $242 in payments.

Month	Previous Balance	Finance Charge 1.5% Monthly	New Purchases	Current Balance	Payment 2% of Balance	Final Balance
January	$1,500.00	$22.50	$150.00	$1,672.50	$33.00	$1,639.50
February	$1,639.50	$24.59	$150.00	$1,814.09	$36.00	$1,778.09
March	$1,778.09	$26.67	$150.00	$1,954.76	$39.00	$1,915.76
April	$1,915.76	$28.74	$150.00	$2,094.50	$42.00	$2,052.50
May	$2,052.50	$30.79	$150.00	$2,233.29	$45.00	$2,188.29
June	$2,188.29	$32.82	$150.00	$2,371.11	$47.00	$2,324.11
Total					$242.00	

Figure 5.2
Making minimum credit card payments and new purchases.

Notice that even though Nicolai has made almost $250 in payments, his balance is almost $1,000 more than his beginning balance in January.

Even if the credit card holder stops using the card to make new purchases, making the minimum payment each month only puts a small dent in the balance because finance charges accrue each month.

Example: From the previous example, what if Nicolai stops using his credit card for new purchases and continues to pay only the minimum payment? What will his balance be after six more months? What percent of his payments went to pay off the balance of the card?

Solution: Continue the table from the previous example, but with no new purchases (see Figure 5.3). To find the percent of payments that went to pay off the balance, divide the payments by the difference between July's previous balance and December's ending balance.

Month	Previous Balance	Finance Charge 1.5% Monthly	New Purchases	Current Balance	Payment 2% of Balance	Final Balance
July	$2,324.11	$34.86	$0.00	$2,358.97	$47.00	$2,311.97
August	$2,311.97	$34.68	$0.00	$2,346.65	$47.00	$2,299.65
September	$2,299.65	$34.49	$0.00	$2,334.14	$47.00	$2,287.14
October	$2,287.14	$34.31	$0.00	$2,321.45	$46.00	$2,275.45
November	$2,275.45	$34.13	$0.00	$2,309.58	$46.00	$2,263.58
December	$2,263.58	$33.95	$0.00	$2,297.53	$46.00	$2,251.53
Total					$279.00	

Figure 5.3
Making minimum credit card payments and no new purchases.

Amount of balance paid off in 6 months = $2,324.11 − $2,251.53 = $72.58

Amount paid in 6 months = $279

Percent of the balance paid = $\dfrac{\$72.58}{\$279} = 0.26 = 26\%$

After 6 more months, Nicolai's balance is $2,251.53. Only 26% of his payments went to paying off the balance.

If Nicolai continues to pay the minimum balance and makes no new purchases, it will take 26 years and 1 month more to pay off the balance of the credit card. In that time, he will have paid $5,147.67 more in finance charges.

PRACTICE PROBLEMS

5.13 Nicolai is going to start making $300 payments on his credit card and making no new purchases. What will his balance be in six more months? Begin with the December final balance from the previous example.

5.14 What percent of his payments will go to paying off the balance?

Table 5.3 shows the time required to pay off a credit card with a 15% APR making only a minimum payment of 1% plus current finance charges and making no new purchases.

Table 5.3 Minimum Credit Card Payments on a 15% APR*

Beginning Balance	Time to Pay Off	Interest Paid	Total Paid
$ 500	3 years, 8 months	$ 150.87	$ 650.87
$ 1,000	8 years, 10 months	$ 729.14	$ 1,729.14
$ 2,000	14 years, 7 months	$ 1,979.25	$ 3,979.25
$ 4,000	20 years, 4 months	$ 4,479.24	$ 8,479.24
$ 5,000	22 years, 2 months	$ 5,729.21	$10,729.21
$10,000	27 years, 11 months	$11,979.29	$21,979.29

*Minimum payments are 1% of balance plus current finance charge. Calculations assume that no new purchases are made on the credit card.

To help educate consumers, the Credit Card Accountability Responsibility and Disclosure Act requires that credit card companies disclose how long it would take to pay off the balance on the credit card if only the minimum payment was made with no new purchases, as well as what payment is required in order to pay off the balance in three years or less.

> **TIP**
>
> Remember, it is possible to use a credit card and pay ZERO finance charges. Use a card with a grace period, pay off your entire balance each month, and do not take cash advances or make balance transfers.

Carrying a high debt, missing payments, and only making minimum payments can affect your *credit score*. Your credit score is a number generated by credit bureaus after they gather information about your credit accounts. With a higher score, you are considered less of a credit risk to lenders and can often obtain credit at lower interest rates. You have the right to receive copies of your credit reports for free once per year and more often under certain conditions.

Assessing Debt

One way to assess debt is to calculate your *debt-to-income ratio*. This ratio shows the percentage of your gross income that is used for debt payments. To find your debt-to-income ratio, find the sum of your monthly debt payments, including child support, auto loans, credit card minimum payments, house payments, and other money spent for housing, such as insurance and taxes. Divide the sum of the debt by your gross income.

$$\text{Debt-To-Income Ratio} = \frac{\text{Monthly Debt Payments}}{\text{Gross Income}}$$

Once you have calculated your debt, you can compare your debt-to-income ratio to this general standard:

- 36% or less: Healthy debt load for most people.
- 37% − 42%: Start reducing debt.
- 43% − 49%: Likely in financial trouble.
- 50% or more: Dangerous financial position.

Example: Angelina has a gross income of $2,000 per month. She pays $650 per month for housing, $200 a month for an auto loan, and $30 per month on her credit card. Find Angelina's debt-to-income ratio and evaluate her debt load based on the standard above.

Solution: Add the monthly debt payments. Divide by the gross income.

Monthly debt payments = $650 + $200 + $30 = $880

$$\text{Debt-to-income ratio} = \frac{\$880}{\$2,000} = 0.44 = 44\%$$

Angelina is likely in financial trouble and should aggressively begin to reduce her debt.

TIP

If you find you are having financial difficulty, start today to get your debt under control.

- Evaluate wants versus needs. Eliminate wants and pay cash for needs.
- Pay more than the minimum payment on credit cards.
- Transfer high interest credit card debt to lower interest accounts.
- Pay off debt with the highest interest rate first.
- Research debt-management counseling.
- Consider consolidating debt into a single low-interest loan and using no more credit.

Chapter

6

Loans

A *loan* is an amount of money that is borrowed and repaid, with interest, on a specified repayment schedule. A loan may be an appropriate choice for an individual or a business to pay for an expensive item and spread out the cost over time.

6.1 Promissory Notes

A *promissory note* is a written agreement to repay a sum of money that you have borrowed. The note should include the terms of the loan, including when the money will be paid and how much interest is due. It may be a simple document like the one in Figure 6.1, or a lengthier document, such as the promissory note on a home loan.

97

```
$___2,500_____Canton, TX_____January 15___20 --

_____Two years_____AFTER DATE I PROMISE TO PAY TO

THE ORDER OF_____Sinclair Bank_____

_____Two Thousand Five Hundred_____DOLLARS

PAYABLE AT_____Sinclair Bank_____

VALUE RECEIVED WITH INTEREST AT __10__%

NO _6389_ DUE____January 15___20 --___Susan Shepherd
```

Figure 6.1
Sample promissory note.

The amount of money borrowed is the *face value* or the *principal* of the note. The *date* of the note is the day that note was signed. The *maturity date* or *due date* is the date when the money must be repaid. The difference between the date of the note and the maturity date is the *time* of the note. The *interest rate* will be stated as an annual interest rate, even if the note is for a time shorter than a year.

To calculate interest on a note that has a single payment due at the maturity date, use the simple interest formula.

$$\text{Interest} = \text{Principal} \times \text{Rate} \times \text{Time}$$

$$I = P \times R \times T$$

Example 1: Find the amount due on the maturity date of the promissory note shown in Figure 6.1.

Solution: Identify the principal, rate and time. Use the simple interest formula and multiply to find the interest. Add the interest to the principal to find the amount due.

$P = \$2,500, R = 10\% = 0.10, T = 2$ years

$I = \$2,500 \times 0.10 \times 2 = \$500$

Amount due $= \$2,500 + \$500 = \$3,000$

On the due date, \$3,000 is due.

PRACTICE PROBLEMS

6.1 Find the amount of interest due on a four-year single payment
promissory note for $8,000 at 5% annual interest.

6.2 Find the amount due at maturity on a one-year single payment
promissory note for $1,000 at 6.5% annual interest.

Exact and Ordinary Interest Methods

Some notes are written for periods shorter than a year. The time on these
notes will typically be expressed in days. A lender may use the exact
interest method or the ordinary interest method to calculate the interest.

Exact interest uses a 365-day year. To calculate exact interest, express the
days as a fraction of a year, using 365 as the number of days in the year.
So, 85 days is expressed as $\frac{85}{365}$ of a year.

Ordinary interest, or *banker's interest method*, uses a 360-day year.
Although no year really has 360 days, historically bankers used a 360-day
year to simplify calculations that were done by hand. Using ordinary
interest, 85 days is expressed as $\frac{85}{360}$ of a year.

Example: Chicha borrows $3,000 for 60 days at 5% interest. Calculate
the interest using exact interest and ordinary interest.

Solution: Identify the principal, rate, and time for exact and ordinary
interest. Multiply to find the interest using each method.

$P = \$3,000, R = 5\% = 0.05$

$T = \dfrac{60}{365}$ for exact interest, $\dfrac{60}{360}$ for ordinary interest

Exact interest $= \$3,000 \times 0.05 \times \dfrac{60}{365} = \$24.66$

Ordinary interest $= \$3,000 \times 0.05 \times \dfrac{60}{360} = \$25.00$

Exact interest is $24.66, and ordinary interest is $25.

Ordinary interest lends itself to calculating interest quickly using fractions. For the previous example, the ordinary interest could have been calculated with the following fractions.

$$\text{Ordinary interest} = \$ \overset{25}{\underset{20}{\cancel{3,000}}} \times \frac{\overset{1}{\cancel{3}}}{\underset{1}{\cancel{100}}} \times \frac{\overset{1}{\cancel{60}}}{\underset{1}{\cancel{360}}} = \$25$$

PRACTICE PROBLEMS

6.3 Calculate the exact interest on $4,000 for 30 days at 3%.

6.4 How much is due on a 60-day note for $1,500 at 4.5% ordinary interest?

Daily Interest Factor

Sometimes a lender will tell you the *daily interest factor* on a note. The daily interest factor is simply the amount of interest accumulating per day. The daily interest factor is calculated by multiplying the principal by the annual interest rate, and then dividing by the number of days in a year. Use 365 for exact interest and 360 for ordinary interest.

$$\text{Daily Interest Factor} = \frac{\text{Principal} \times \text{Rate}}{\text{Number of Days in a Year}}$$

Example: Find the daily interest factor on $1,200 borrowed at 6% exact interest. Round to the nearest ten-thousandth.

Solution: Exact interest uses a 365-day year. Multiply the principal by the interest rate expressed as a decimal, and then divide by 365.

Interest rate $= 6\% = 0.06$

$$\text{Daily interest factor} = \frac{\$1,200 \times 0.06}{365} = \$0.1973$$

The daily interest factor is $0.1973.

The loan is accumulating interest at the rate of almost 20 cents per day.

If a lender tells you the daily interest factor, you can quickly calculate the amount of interest due for any number of days. Simply multiply the daily interest factor by the number of days.

$$\text{Interest} = \text{Daily Interest Factor} \times \text{Number of Days}$$

Example: If the daily interest factor is $0.1973, what amount of interest is due in 60 days?

Solution: Multiply the daily interest factor by 60, the number of days.

Interest = $0.1973 \times 60 = $11.84

The interest due in 60 days is $11.84.

Practice Problems

6.5   Find the daily interest factor on a 90-day note for $2,000 at 6% exact interest.

6.6   If the daily interest factor on a note is $0.0836, how much interest is due in 30 days?

6.2 Installment Loans

An *installment loan* is a loan that is repaid with a fixed number of periodic equal-sized payments.

Merchant Installment Plans

Many merchants offer installment financing, called an *installment plan.* An installment plan will typically require a *down payment,* an upfront payment made toward the cost of the merchandise, in addition to the periodic payments. The price you pay for an item with an installment plan will be higher than the price you would pay if you purchased the item for cash. The additional money paid on the installment plan is the finance charge.

Example: Bob purchases living room furniture on an installment plan. He makes a down payment of $200 and 24 monthly payments of $55. The regular cash price of the furniture is $1,100. What is the installment price? By what percent is the installment price greater than the cash price? Round to the nearest percent.

Solution: To find the installment price, multiply the monthly payment by the number of payments, and then add the down payment. Subtract the cash price from the installment price to find the finance charge. Divide the finance charge by the cash price to find the percent the installment price is greater than the cash price.

Total of monthly payments = 24 × $55 = $1,320

Installment price = $1,320 + $200 = $1,520

Finance charge = $1,520 − $1,100 = $420

Percent increase = $420 ÷ $1,100 = 0.38 = 38%

The installment price is $1,520, and is 38% greater than the cash price.

Other merchants may advertise low payments, but make sure you know the amount of the payment before you make the purchase.

Example: A TV is advertised with an installment price of $850 with a down payment of $100 and 12 easy payments. How much are the monthly payments?

Solution: Subtract the down payment from the installment price to find the total amount financed. Divide by the number of payments to find the amount of each payment.

Amount financed = $850 − $100 = $750

Monthly payments = $750 ÷ 12 = $62.50

The monthly payments are $62.50.

PRACTICE PROBLEMS

6.7   Cecilia purchased a new cell phone on an installment plan. The cash price of the phone is $200. She makes a down payment of $40 and six monthly payments of $30. What is the installment price of the phone, and what percent greater is it than the cash price?

6.8   Abul sees an advertisement for a computer that can be paid for on an installment plan with a $50 down payment and 24 monthly payments. If the installment price of the computer is $650, how much are the monthly payments?

Bank Installment Loans

Instead of paying a merchant for an installment plan, you can obtain an installment loan from a bank or credit union. Typically, the interest rate for an installment loan will be lower than the interest rate on a merchant's installment plan.

In a *simple interest installment loan*, you pay interest each month on the amount of the loan that remains. Each month your payment is applied first to the interest due that month, and the remainder of the payment is used to reduce the principal.

Table 6.1 shows a loan repayment schedule on a one year $1,000 loan at 9% annual interest.

Table 6.1 Loan Repayment Schedule

Month	Monthly Payment	Interest Payment	Applied to Principal	Balance
1	$87.45	$7.50	$79.95	$920.05
2	$87.45	$6.90	$80.55	$839.50
3	$87.45	$6.30	$81.16	$758.34
4	$87.45	$5.69	$81.76	$676.58
5	$87.45	$5.07	$82.38	$594.20
6	$87.45	$4.46	$82.99	$511.21
7	$87.45	$3.83	$83.62	$427.59
8	$87.45	$3.21	$84.24	$343.34
9	$87.45	$2.58	$84.88	$258.47
10	$87.45	$1.94	$85.51	$172.95
11	$87.45	$1.30	$86.15	$ 86.80
12	$87.45	$0.65	$86.80	$ 0.00

Notice that the monthly payments are the same. The interest payment decreases, and the amount of the payment applied to principal increases each month as the balance declines.

If you know the balance of a loan, the annual interest rate, and the amount of each payment, you can calculate the amount of interest and the amount of payment applied to the principal by using the simple interest formula.

$$\text{Interest} = \text{Principal} \times \text{Rate} \times \text{Time}$$

$$I = P \times R \times T$$

Example: Simeon borrowed $2,000 with a two-year simple interest installment loan at 6% annual interest. The monthly payment is $88.64. Find the amount of interest, the amount applied to the principal, and the new balance for the first monthly payment.

Solution: Find the monthly interest rate by dividing the annual interest rate by 12. Use the simple interest formula to calculate the interest due in the first month. Subtract the interest from the monthly payment to find the amount applied to the principal. Subtract the amount applied to the principal from the balance to find the new balance.

Monthly interest rate = 6% ÷ 12 = 0.5% = 0.005

$I = P \times R \times T = \$2,000 \times 0.005 \times 1 = \$10$

Amount applied to principal = $88.64 − $10 = $78.64

New balance = $2,000 − $78.64 = $1,921.36

The amount of interest in the first month is $10, the amount applied to the principal is $78.64, and the new balance is $1,921.36.

PRACTICE PROBLEMS

6.9   Tanner borrowed $500 on a one-year simple interest loan at 12% annual interest. His monthly payment is $44.42. Find the amount of interest, the amount applied to the principal, and the new balance for the first monthly payment.

6.10   Jennifer signed a $4,000, nine-month simple interest installment loan at 9% annual interest. The monthly payments are $461.28. Find the interest, the amount applied to principal, and the new balance for the first two months.

6.3 Annual Percentage Rate

Before taking out a loan, it is important to consider the interest rate. A lending institution will typically quote an annual interest rate and the *annual percentage rate*, or APR, for a loan. The annual interest rate is the rate that is used to calculate the payments for the loan. The APR is the yearly interest rate including interest and fees. Lending institutions are required to disclose the APR of a loan.

To find the APR for a loan payable in a lump sum at the maturity date, divide the finance charge for one year, including interest and fees, by the principal, or the amount financed. If the finance charge is for less than one year, calculate how much the finance charge would be for a year.

$$\text{APR for Single Payment Loan} = \frac{\text{Finance Charge for One Year}}{\text{Principal}}$$

Payday loans are short-term loans, typically offered by check-cashing businesses. These loans are often for just a week or two, or until your next payday.

Example: A payday loan lender offers to loan you $200 for two weeks with a $50 finance charge. What APR is the lender offering?

Solution: The quoted interest is for two weeks, so multiply the finance charge by 26 to find the amount of finance charge for 52 weeks, or one year. Divide the interest by the principal to find the interest rate.

Finance charge = $50 × 26 = $1,300

$$\text{APR} = \frac{\$1,300}{\$200} = 6.5 = 650\%$$

The APR for the payday loan is 650%.

TIP

Individuals who take out a payday loan often find that they are still unable to meet their financial obligations on the next payday, and another loan is needed. In fact, only 2% of payday loan borrowers use only a single payday loan to meet their financial obligations. The rest require multiple loans, usually at triple-digit interest rates.

PRACTICE PROBLEMS

6.11 A friend offers to loan you $400 for one month if you pay him back $415. What APR is the friend charging you?

6.12 A check-cashing business charges $50 for a $200 loan due in one week. What is the APR?

The interest rate for an installment loan takes into account the fact that, since you are paying the money back incrementally, you did not have use of the entire amount of the loan for the entire loan period.

Consider if you borrow $1,000 for one year and make a single payment of $1,050 at the end of the year. You paid $50 in interest, and your APR $= \frac{\$50}{\$1,000} =$ 5%, however, if you pay $50 in interest on a $1,000 installment loan paid monthly over one year, your APR is 9.2%.

You can use this formula to estimate the APR for an installment loan.

$$\text{APR} = \frac{2 \times \text{Annual \# of Payments} \times \text{Finance Charge}}{(\text{Total \# of Payments} + 1) \times \text{Principal}}$$

Example: A car dealer offers to finance $7,000 for 36 months with monthly payments of $225. What is the APR to the nearest tenth of a percent?

Solution: Calculate the amount of interest by multiplying the monthly payment by 36 and subtracting $7,000. Substitute the values into the APR formula and evaluate.

Total payments $= \$225 \times 36 = \$8,100$

Interest $= \$8,100 - \$7,000 = \$1,100$

Annual \# of payments $= 12$

Total \# of payments $= 36$

Principal $= \$7,000$

$$\text{APR} = \frac{2 \times 12 \times \$1,100}{(36 + 1) \times \$7,000}$$

$$\text{APR} = \frac{\$26,400}{\$259,000} = 10.2\%$$

The car dealer is offering a loan with an APR of 10.2%.

If there are no additional fees to the loan, then the annual interest rate and the APR will be the same. If there are fees, then the APR will be higher than the annual interest rate. All other things being equal, you want the loan with the lowest APR.

Example: Consider the loan in previous example. If the loan has a $100 processing fee, what is the APR to the nearest tenth of a percent?

Solution: Substitute the values into the APR formula and evaluate.

Annual # of payments = 12

Finance charge = $1,100 + $100 = $1,200

Total number of payments = 36

Principal = $7,000

$$APR = \frac{2 \times 12 \times \$1,200}{(36 + 1) \times \$7,000}$$

$$APR = \frac{\$28,800}{\$259,000} = 11.1\%$$

The APR increases from 10.2% to 11.1% with the fee.

PRACTICE PROBLEMS

6.13 The finance company for an auto dealer offers to finance $5,000 for 72 months, with a monthly payment of $102.58. What APR is the company offering?

6.14 If the finance company has a processing fee of $85, what is the APR?

6.4 Early Loan Repayments

To pay a simple interest installment loan off early, you pay the current month's interest charge and the remaining balance on the loan as your final payment.

Example: Marty has a 24-month, $5,000 loan at 6% annual interest. His payments are $221.60. After making payments on the loan for 11 months, his balance is $2,782.53. He receives a bonus from work and decides to pay off the loan with his next payment. How much will his final payment be?

Solution: Divide the annual interest by 12 to find the monthly interest. Multiply the monthly interest rate by the current balance to find the interest due this month. Add the interest due this month to his balance to find his final payment.

Monthly interest rate = 6% ÷ 12 = 0.5% = 0.005

Interest = $2,782.53 × 0.005 = $13.91

Final payment = $2,782.53 + $13.91 = $2,796.44

Marty's final payment will be $2,796.44.

A benefit to paying off a simple interest installment loan is that you pay less interest.

Example: In the previous example, how much interest does Marty save by paying off the loan early?

Solution: Multiply the monthly payment by 24 to find out how much Marty would have paid if he had paid out the loan with the scheduled payments. Multiply the monthly payment by 11 to find out how much Marty paid in the first 11 months of the loan. Add that amount to the final payment to find out how much Marty paid in total for the loan. Subtract that amount from the amount he would have paid if he had made all 24 payments.

Total for 24 payments = $221.60 × 24 = $5,318.40

Amount paid in 11 months = $221.60 × 11 = $2,437.60

Total paid including final payment = $2,437.60 + $2,796.44 = $5,234.04

Interest saved = $5,318.40 − $5,234.04 = $84.36

Marty saved $84.36 in interest by paying off the loan early.

PRACTICE PROBLEMS

6.15 June financed her $25,000 auto loan at 12% for 60 months. Her monthly payment is $556.11. After paying on the loan for one year, her balance is $21,117.76, but she decides she wants to sell the car and get a vehicle that is more affordable. How much does she need to sell the car for in order to be able to make her final payment with the 13th payment?

6.16 How much will June save in interest by selling the car and paying the loan off early?

Some loans have a *prepayment penalty*, which is either a set amount, a percentage of the loan, or a certain number of months' interest that you will be charged in addition to your final payment if you pay the loan off early.

TIP

A Rule of 78 loan, also called a *precomputed interest loan*, penalizes you for prepayment because at the beginning of the loan you pay more than just the current month's interest. If you pay the loan off early, you will have paid more interest than a simple interest installment loan. Rule of 78 loans are illegal for loan terms longer than five years nationwide and are illegal for any term in some states.

Be certain of your plans before you sign any loan that penalizes you for prepayment.

Chapter

7

Auto and Home Ownership

H ome and auto purchases are the two biggest purchases most people will make in their lifetime. In addition to the initial cost to purchase or finance a home or car, there are many other ongoing costs associated with these purchases, such as insurance, routine maintenance, and repairs.

7.1 Mortgages

A *mortgage* is a loan secured by a house and the land it is on. In other words, if you do not pay your mortgage, your lender has the legal right to *foreclose* on your home and land and take your property to sell it and pay the debt that you owe.

A mortgage is like other loans. It has a promissory note with an interest rate and the terms by which it will be repaid. Mortgages are typically for 15 or 30 years. There are several kinds of mortgages, but the most common are fixed-rate and variable-rate mortgages. In a *fixed-rate mortgage*, the interest rate does not change over the life of the loan. In a *variable-rate mortgage*, the interest rate can change. Variable-rate mortgages will offer lower interest rates than fixed-rate mortgages, but it is possible that the interest rate will increase over the life of the loan.

Qualifying for a Mortgage

To determine whether you qualify for a mortgage, your lender will evaluate your creditworthiness. Important factors include your income, your credit score, and your debt-to-income ratio. Many lenders apply the 28/36 standard for debt-to-income ratio.

The 28/36 debt-to-income standard is as follows:

- No more than 28% of gross income for housing, including mortgage payment, insurance and taxes.
- No more than 36% of gross income for housing and other debt.

Example: If your gross income is $2,500 per month, what is the maximum amount that a lender will allow for housing and other debt using the 28/36 debt-to-income standard?

Solution: Multiply the gross income by 28% to find the amount allowed for housing. Multiply the gross income by 36% to determine how much is allowed for housing and other debt.

Maximum allowed for housing = $2,500 × 28% = $2,500 × 0.28 = $700

Maximum allowed for housing and other debt = $2,500 × 36% = $2,500 × 0.36 = $900

Using the 28/36 standard, a lender would allow $700 for housing costs and $900 for housing and other debt.

PRACTICE PROBLEMS

7.1 If your gross income is $3,000 per month, what is the maximum amount that a lender will allow for housing using the 28/36 debt-to-income ratio?

7.2 By the 28/36 standard, can you afford a home that will have housing costs of $1,400 per month if you make $5,000 per month and have other debt obligations of $600 per month?

Down Payment and Closing Costs

When you purchase a home with a mortgage, your lender should require a *down payment*, money that you pay upfront toward the cost of the home. Your lender may require that you pay a certain percentage, some as much as 30%, although you can make a bigger down payment than your lender requires. Since your down payment is going toward the cost of the house, the higher your down payment, the less money you have to borrow and the less you will pay monthly and in interest.

To find the amount you need to borrow, subtract your down payment from the purchase price.

Mortgage Amount = Purchase Price − Down Payment

In addition to the down payment, you will need money for the *closing costs*. These are costs associated with the loan and the purchase of the house. Typical closing costs are legal, appraisal, inspection and recording fees, title insurance, loan application fees, land surveys, prepaid taxes, and prepaid interest charges, or *points*. Closing costs will vary by loan, so make sure you understand the closing costs for the loans you are considering. To calculate how much cash you need to purchase a home, add the down payment and the closing costs.

Cash Needed to Purchase Home = Down Payment + Closing Costs

Example: June is purchasing a home for $85,000. Her lender requires a 20% down payment and $2,000 in closing costs. What is the amount of her mortgage? How much cash will she need to purchase the home?

Solution: Find the amount of the down payment by multiplying the purchase price by 20%. Subtract the down payment from the purchase price to find the amount of the mortgage. Add the down payment and the closing costs to find the amount of cash needed.

Down payment = $85,000 × 20% = $85,000 × 0.20 = $17,000

Amount of mortgage = $85,000 − $17,000 = $68,000

Cash needed = $17,000 + $2,000 = $19,000

June will have a $68,000 mortgage and needs $19,000 at closing to purchase the house.

Practice Problems

7.3   Lucas is purchasing a home for $125,000. He is making a 30% down payment and estimates the closing costs at $4,000. What is the amount of the mortgage?

7.4   Jan is planning to purchase a home for $75,000. She will make a 15% down payment. Closing costs are estimated at 2% of the mortgage. How much cash will she need at closing?

Monthly Payments and Interest

Most mortgages are paid with equal-sized monthly payments. Just like a simple interest installment loan, each month the payment pays the current month's interest and part of the principal. Most lenders will accept additional payments to decrease the principal of the loan and pay it off faster with less interest.

If you know the monthly payment and the number of years of the loan, you can calculate how much interest will be paid over the life of the loan.

Example: Oswald wants to purchase a house that costs $100,000. His mortgage lender requires a 15% down payment for a 30-year loan. His monthly mortgage payment will be $509.62. How much will Oswald pay back over 30 years? How much will he pay in interest?

Solution: Multiply the purchase price of the house by 15% to determine his down payment. Subtract the down payment from the purchase price to find out the amount of his mortgage. Multiply $509.62 by 360 (the number of months in 30 years, or 12 × 30) to find out how much he pays back. To find the interest paid, subtract the amount he borrowed from the amount he paid back.

Down payment = $100,000 × 15% = $100,000 × 0.15 = $15,000

Amount of mortgage = $100,000 − $15,000 = $85,000

Amount paid back = $509.62 × 360 = $183,463.20

Interest = $183,463.20 − $85,000 = $98,463.20

Oswald pays back $183,463.20, including $98,463.20 of interest.

PRACTICE PROBLEMS

7.5 Kim purchases a home and takes out a 30-year mortgage for $65,000. Her monthly mortgage payment is $389.71. How much will she pay back over the life of the loan?

7.6 Cary is purchasing a home for $200,000. She is making a 20% down payment and will sign a 15-year mortgage. Her monthly mortgage payment will be $1,350.17. How much will Cary pay over the life of the mortgage? How much will she pay in interest?

A mortgage lender may offer or require that you open an *escrow account*. Each month, in addition to your monthly mortgage payment, you will pay an additional amount to the lender, which is held in the escrow account. The money in the escrow account is used to pay property taxes and homeowner's insurance premiums.

Refinancing

If you have a fixed-rate mortgage and the economy shifts so that mortgage interest rates have dropped, you may want to *refinance* your mortgage. Refinancing a mortgage means to obtain a new mortgage. When you obtain a new mortgage, the money borrowed for the new mortgage is used to pay off the old mortgage.

Since you will usually have to pay closing costs, you don't actually save money until you reach the *breakeven point*, or the number of months it takes for your savings in monthly payments to be greater than or equal to the amount it costs to refinance.

- Breakeven Point for Refinancing $= \dfrac{\text{Total Cost of Refinancing}}{\text{Savings Each Month}}$

- Savings Each Month = Old Monthly Payment − New Monthly Payment

Example: Sonia has paid on a 30-year loan for 10 years. Her interest rate is 7%, and her current monthly payment is $665.30. She can refinance the current mortgage and get a 20-year mortgage at 5%, and her new payment would be $556.33. If the closing costs to refinance are $1,500, how many months would it take to reach or pass the breakeven point?

Solution: Subtract the new payment from the old to find the savings each month. Divide the closing costs by the savings each month.

Savings each month = $665.30 − $556.33 = $108.97

$$\text{Breakeven point} = \frac{\$1,500}{\$108.97} = 13.8$$

It would take approximately 14 months to pass the breakeven point.

PRACTICE PROBLEMS

7.7 If your current mortgage payment is $875, and you can refinance for $2,000 and have a $750 mortgage payment, how many months will it take to reach or pass the breakeven point?

7.8 Jan has a monthly mortgage payment of $580. She can refinance and have a mortgage payment of $500 if she pays $1,500 in closing costs. If she plans to change jobs and move in the next 18 months, should she refinance? Why or why not?

TIP

If you refinance for a period that is greater than the time you have left on your current mortgage, you can have an even lower monthly payment, but you may end up paying more in total interest. Beware of refinancing just to lower your monthly payments.

7.2 Property Taxes

Property tax, also called *real estate tax*, is money that is collected by a tax district, often a city, town, or county, based on the *assessed value* of property. Property tax money is often used for police and fire departments, schools, and roads.

Tax rates are often expressed as a dollar amount per hundred or per thousand. To find the amount of property tax owed, divide the assessed value by $100 or $1,000, and then multiply by the tax rate.

Example: If the tax rate in Wise County is $5.18 per $100, how much property tax is owed on a property with an assessed value of $120,000?

Solution: Divide the assessed value by $100, and then multiply by the tax rate.

$120,000 ÷ $100 = 1,200

1,200 × $5.18 = $6,216

The property tax owed is $6,216.

Example: How much is owed on the property in the previous example if the tax rate is $5.18 per $1,000?

Solution: Divide the assessed value by $1,000, and then multiply by the tax rate.

$120,000 ÷ $1,000 = 120

120 × $5.18 = $621.60

The property tax owed is $621.60.

Some tax districts use a tax rate in *mills*. A mill is one-thousandth of a dollar. To change the mills to dollars, divide the mills by 1,000. Multiply the tax rate in dollars by the assessed value to find the property tax.

Example: If the tax rate in the city of Hooperville is 55 mills, find the property tax due if the assessed value is $120,000.

Solution: Divide the mills by 1,000 to change the tax rate to dollars. Multiply by the assessed value to find the property tax.

55 mills ÷ 1,000 = $0.055

Property tax = $120,000 × $0.055 = $6,600

The property tax owed is $6,600.

Other tax districts use a tax rate in cents per dollar. To change cents into dollars, divide by 100. Multiply the tax rate in dollars by the assessed value to find the property tax.

Example: How much is owed in property taxes on the property in the previous example if the tax rate is 7.3 cents per dollar?

Solution: Divide the tax rate by 100. Multiply by the assessed value.

Tax rate = 7.3 cents ÷ 100 = $0.073

Property tax = $120,000 × $0.073 = $8,760

The property tax owed is $8,760.

TIP

Equivalent tax rates:
5 cents per dollar = 50 mills = $5 per hundred = $50 per thousand

---

Some tax districts allow exemptions that reduce the amount of the assessed value that is taxed. For example, properties owned by senior citizens or properties that are occupied by owners instead of renters may qualify for exemptions in certain tax districts.

Practice Problems

7.9   Find the tax owed on a property with an assessed value of $85,000 if the tax rate is $3.45 per hundred.

7.10   Find the tax owed on a property with an assessed value of $85,000 if the tax rate is $9.43 per thousand.

7.11   Find the tax owed on a property with an assessed value of $85,000 if the tax rate is 48 mills.

7.12   Find the tax owed on a property with an assessed value of $85,000 if the tax rate is 3.2 cents per dollar.

7.3 Property Insurance

Property insurance covers you and your property against certain losses and risks. Property insurance purchased specifically for your home is called *homeowner's insurance*. The amount for which you insure your property is called the *face value* of the policy. The company you purchase insurance from is called the *insurer*. You pay the insurer a *premium*, typically a yearly amount, for the insurance. The amount of your premium depends on a variety of things, including the age of your home and its roof, the location, and the face value of the policy.

Property insurance rates are usually based on $100 units of insurance. Insurance premiums are rounded to the nearest dollar.

Example: Sharon insured her home for $95,000 at an annual rate of $0.63 per $100. What is the insurance premium for one year?

Solution: Divide the face value of the policy by $100 to find how many $100 units are needed. Multiply the number of $100 units by the annual rate.

$95,000 ÷ $100 = 950

950 × $0.63 = $598.50 = $599, rounded to the nearest dollar

The annual premium is $599.

PRACTICE PROBLEM

7.13 Jack has a home insured for $55,000. The premium is $0.72 per $100. How much is the annual premium?

Making an Insurance Claim

If you have a loss that is covered by your insurance policy, you must file a claim with the insurer within a specified time period. The insurance company will send an adjuster to your home to evaluate the dollar amount of your loss. With a basic policy, your insurer will pay you the amount of your loss, up to the face value of your policy, less the deductible. The *deductible* is the initial increment of loss that you are responsible to pay. A policy with a higher deductible will have a lower premium.

Example: A home that is insured for $66,000, suffers a loss of $25,000. The policy has a $1,000 deductible. How much will the insurance company pay?

Solution: Verify that the loss is less than the face value of the policy. If so, subtract the deductible from the loss.

$25,000 is less than $66,000.

$25,000 − $1,000 = $24,000

The insurance company will pay $24,000.

To save money on premiums, some property owners will insure their property for less than the value of the property. Many property insurance policies will have a *coinsurance clause*, which imposes a penalty if there is a claim and the property owner does not carry at least the required amount of insurance on the property. The penalty reduces the amount paid on a claim. Typically, a coinsurance clause will require the property owner to carry insurance to cover at least 80% of the property's value.

$$\text{Amount of Claim Paid} = \frac{\text{Face Value of Policy}}{\text{Required Amount of Insurance}} \times \text{Amount of Loss}$$

Example: Frances owns a home valued at $120,000. Her homeowner's insurance policy has an 80% coinsurance clause. She insures the home for $90,000. She files an insurance claim for $8,000. How much of the claim will the insurance company pay?

Solution: Multiply the value of the home by 80% to determine the amount of insurance required. Divide the face value by the required amount of insurance and multiply by the amount of loss.

Percent of insurance required = 80% = 0.80

Required amount of insurance = $120,000 × 0.80 = $96,000

$$\text{Amount of Claim Paid} = \frac{\$90,000}{\$96,000} \times \$8,000 = \$7,500$$

The insurance company will pay $7,500 of the claim.

PRACTICE PROBLEMS

7.14 A home that is insured for its full value of $95,000 suffers fire damage of $6,000. If the policy has a $500 deductible, how much will the insurance company pay on the claim?

7.15 Hamil insures his $145,000 home for $110,000 with a homeowner's policy that has an 80% coinsurance clause. The house suffers damage totaling $50,000. How much will the insurance company pay on the claim?

In addition to mortgage payments, taxes and insurance, there are many other costs associated with owning a home, including the cost to furnish the home with furniture and appliances, routine maintenance, and repairs.

7.4 Buying a Car

There are many costs and adjustments that impact the total cost of a car. If the car is purchased from a car dealer, sales tax is computed on and added to the purchase price.

In addition, a dealer will add registration fees and other non-taxable costs, such as an extended warranty. If you purchase a car from an individual, in most states you will pay sales tax and registration fees when you register the car.

A dealer may also offer a rebate that reduces the purchase price. The *delivered price* of the car is the cost after taxes, fees, and other costs are added, and any rebate is subtracted. If you finance the car, you will probably make a down payment. The *balance due* on the car is the amount left after the down payment is subtracted from the delivered price.

- Delivered Price = Purchase Price + Sales Tax + Registration Fees + Non-taxable Items − Rebate
- Balance Due = Delivered Price − Down Payment

Example: Sylvia negotiates a purchase price of $18,500 on a new car. There is a 4% sales tax and registration fees of $239. She receives a rebate of $500 from the manufacturer. She makes a down payment of $1,000. What are the delivered price and the balance due?

Solution: Multiply the purchase price by the sales tax rate to find the sales tax. Add the sales tax and registration fees to the purchase price and subtract the rebate to find the delivered price. Subtract the down payment from the delivered price to find the balance due.

Sales tax rate = 4% = 0.04

Sales tax = $18,500 × 0.04 = $740

Delivered price = $18,500 + $740 + $239 − $500 = $18,979

Balance due = $18,979 − $1,000 = $17,979

The delivered price is $18,979, and the balance due is $17,979.

PRACTICE PROBLEMS

7.16   Josh buys a used car from a car dealer for $6,000. He is charged 5% sales tax and registration fees of $125. What is the delivered price of the vehicle?

7.17   Fern purchases a new car for $23,800. She pays 3.5% sales tax and registration fees of $250. The dealer gives her a rebate of $500, and she purchases a non-taxable extended warranty for $1,000. She makes a down payment of $2,000. What is the balance due?

Most vehicles are financed with simple interest installment loans. To find the total cost of the car, multiply the monthly payment by the number of payments and add the down payment.

Example: Joanna negotiates a purchase price of $12,400 on a new car. The delivered price of the car is $13,800. She makes a $1,500 down payment and finances the balance due. Her payments are $306.09 per month for 48 months. What is the total cost of the car? How much more did she pay for the car by financing the balance due?

Solution: Multiply the monthly payment by 48 to find the total amount paid with the loan. Add the down payment to find the total amount paid for the car. Subtract delivered price from the total amount paid for the car to find the amount paid for financing the vehicle.

Total loan payments = $306.09 × 48 = $14,692.32

Total paid for car = $14,692.32 + $1,500 = $16,192.32

Cost of financing = $16,192.32 − $13,800 = $2,392.32

The total cost of the car is $16,192.32. She paid $2,392.32 more by financing the car with a loan.

PRACTICE PROBLEMS

7.18 Jose purchases a car for a delivered price of $29,400. He makes a $5,000 down payment and finances the balance due with 48 monthly payments of $573.03. What is the total cost of the car?

7.19 In the previous exercise, how much did Jose pay in interest?

Although most vehicle loans are financed with simple interest installment loans, in some states, Rule of 78 loans with terms of five years or less are legal. In a Rule of 78 loan, your early payments go toward paying off a greater percent of the precomputed interest. If you pay off a Rule of 78 loan early, you will pay more in interest than you would if you paid off a simple interest loan with no prepayment penalty early.

7.5 Depreciation

A vehicle loses value over time due to wear and tear and obsolescence. The loss of value is called *depreciation*. The difference between the original value of the vehicle and its resale or trade-in value is the amount of depreciation.

$$\text{Depreciation} = \text{Original Value} - \text{Resale or Trade-in Value}$$

A vehicle depreciates most in its first year, although calculating the average annual depreciation can give a good picture of depreciation over time.

$$\text{Average Annual Depreciation} = \frac{\text{Depreciation}}{\text{Number of Years}}$$

Example: A car purchased new for $25,000 five years ago is now worth $10,100. What are the total depreciation and the average annual depreciation?

Solution: Subtract the current value from the original value to find the total depreciation. Divide the total depreciation by 5 to find the average annual depreciation over the 5 year time period.

Depreciation = $25,000 − $10,100 = $14,900

$$\text{Average annual depreciation} = \frac{\$14,900}{5} = \$2,980$$

The total depreciation in five years is $14,900. The average annual depreciation is $2,980.

The rate of depreciation is the percent of value that is being lost per year. To find the rate of depreciation, divide the average annual depreciation by the original value and change to a percent by multiplying by 100 or moving the decimal point two places to the right.

$$\text{Rate of Depreciation} = \frac{\text{Average Annual Depreciation}}{\text{Original Value}}$$

Example: Find the rate of depreciation rounded to the nearest hundredth of a percent for the car in the previous example.

Solution: Divide the average annual depreciation by $25,000, the original cost of the car. Multiply the result by 100 or move the decimal point two places to the right to change to a percent.

Average annual depreciation from previous example = $2,980

Original value = $25,000

$$\text{Rate of depreciation} = \frac{\$2,980}{\$25,000} = 0.1192 = 11.92\%$$

The rate of depreciation is 11.92%

Although you can calculate the average annual depreciation and the rate of depreciation, neither is an exact measure of the value from year to year. Despite any calculations, the actual value of a vehicle is still the amount that you can sell it for.

PRACTICE PROBLEMS

7.20 A $15,000 car is valued at $8,500 after four years. What are the total depreciation and the average annual depreciation?

7.21 In the previous exercise, what is the rate of depreciation, rounded to the nearest hundredth of a percent?

7.6 Leasing a Car

Instead of buying a car, some people *lease* a car. Leasing a car is similar to renting a car because you use the car for a time, and then you return it. The cost of leasing is based on the depreciation of the car over the time period of the lease plus a finance charge.

The leasing agent and the person leasing the car must negotiate a *lease price*, which is the beginning value of the car, and a *residual value*, which is the estimated depreciated value of the car at the end of the lease. In addition, the APR or *money factor* must be agreed upon. The money factor is the APR divided by 2,400 and is used to calculate the finance charges for the lease.

To find the monthly payments on a lease, first calculate the average monthly depreciation. Subtract the residual value from the lease price and divide by the number of months of the lease.

$$\text{Average Monthly Depreciation} = \frac{\text{Lease Price} - \text{Residual Value}}{\text{Number of Months}}$$

The finance charge is calculated by adding the lease price and the residual value and multiplying by the money factor.

- Money Factor = $\dfrac{\text{APR}}{2{,}400}$

- Finance Charge = (Lease Price + Residual Value) × Money Factor

The monthly lease payment is the sum of the average monthly depreciation and the finance charge.

- Monthly Lease Payment = Average Monthly Depreciation + Finance Charge.

In some states, sales tax will be added to the monthly lease payment. The method for calculating sales tax on a lease varies from state to state.

Example: Raven negotiates a 36-month car lease with a lease price of $20,000 and a residual value of $12,800. The APR is 9%. What is the monthly lease payment?

Solution: Calculate the average monthly depreciation by subtracting the residual value from the lease price and dividing by 36 months. Divide the APR by 2,400 to find the money factor. Calculate the finance charge by adding the lease price and the residual value and multiplying by the money factor. Add the average monthly depreciation and the finance charge to find the monthly lease payment.

$$\text{Average monthly depreciation} = \frac{\$20{,}000 - \$12{,}800}{36} = \$200$$

$$\text{Money Factor} = \frac{9}{2{,}400} = 0.00375$$

Finance charge = ($20,000 + $12,800) × 0.00375 =
$32,800 × 0.00375 = $123

Monthly lease payment = $200 + $123 = $323

The monthly lease payment without sales tax is $323.

PRACTICE PROBLEMS

7.22 What is the monthly lease payment on a car with a lease price of $45,000, a residual value of $20,000 on a five-year lease if the money factor is 0.00392?

7.23 Justin negotiates a lease price of $20,000, and a residual value of $12,000 on a four-year lease. If the APR is 9.5%, what is the monthly lease payment?

Most lease contracts have a mileage allowance. If you drive the car more than the mileage allowance during the term of the lease, you will pay an additional fee, usually expressed as a cost per mile.

Example: Jory leases a car for $350 per month for 48 months and a 48,000 mileage allowance. When she returns the car at the end of the 48 months, she has driven the car 52,500 miles. If the charge is $0.25 per mile for miles over the mileage allowance, what is her total cost to lease the car for 48 months?

Solution: Find the total lease payments made multiplying the monthly lease by 48. Subtract the mileage allowance from the number of miles driven to find the number of excess miles. Multiply the excess miles by the cost per mile. Add the total lease payments and the excess mileage fee to find the total cost.

Total lease payments = $350 × 48 = $16,800

Excess miles = 52,500 − 48,000 = 4,500

Excess mileage fee = 4,500 × $0.25 = $1,125

Total cost to lease = $16,800 + $1,125 = $17,925.

Jory's total cost to lease the car for 48 months is $17,925.

PRACTICE PROBLEMS

7.24   What is the mileage fee for a leased car that is driven 100,000 miles with a mileage allowance of 85,000 and an excess mileage charge of $0.225 per mile?

7.25   Priscilla leases a car for $239 per month for 48 months and a 40,000 mileage allowance. The charge for excess miles is $0.18 per mile. If she turns the car in at the end of the lease period and has driven 45,000 miles, what is her total lease cost?

7.7 Auto Insurance

Like property insurance, auto insurance protects you against loss. There are four basic types of auto insurance:

* *Bodily injury* covers your liability for injury you cause to others.

* *Property damage* covers your liability for injury you cause to others' property.

* *Collision* covers damage you cause to your own vehicle.

* *Comprehensive damage* covers damage or loss to your vehicle from causes other than collisions, such as fire, vandalism, and hail.

Bodily injury and property damage insurance cover your damage to others and together are called *liability insurance*. All states require that you carry at least a minimum amount of liability insurance. For example, a state may have a 25/50/15 minimum liability requirement. The first two numbers refer to bodily injury, and the third number refers to property damage. Coverage of 25/50/15 translates to $25,000 of bodily injury coverage per person, $50,000 per accident, and $15,000 of property damage coverage per accident. Liability insurance does not cover repair and replacement of your own vehicle.

Collision and comprehensive insurance cover the repair and replacement of your vehicle in case of an accident that you cause or other non-collision losses. If you financed your car with a loan, your lender may require you to carry collision and comprehensive insurance. Collision and comprehensive insurance have a deductible, an amount you must pay toward each loss that you claim with the insurance company. The lower the deductible, the higher the premium will be.

In addition to the basic forms of insurance, you will have the option to purchase *uninsured* or *underinsured motorist insurance*. This insurance covers you and your property if the driver who causes a collision is not adequately insured to cover your loss.

Premiums vary greatly and are set considering a variety of factors. Some of those factors include the age and driving record of the drivers, make and model of the vehicle, location, and primary use of the vehicle.

Example: Jenny lives in a state that requires 25/50/25 liability insurance. She plans to carry the minimum liability insurance plus collision and comprehensive insurance with $100 deductibles. Her vehicle is used for driving to work. What is her annual insurance premium based on Figure 7.1?

Sample Annual Car Insurance Premiums

Type of Insurance Coverage	Coverage Limits	Annual Premiums for:		
		Pleasure Use Only	Driving to Work	Business
Bodily Injury	$25/50,000	$20.58	$22.84	$29.71
	50/100,000	30.88	34.27	44.68
	100/300,000	53.95	59.35	79.74
Property Damage	$25,000	$135.80	$150.74	$196.50
	50,000	161.67	179.44	233.92
	100,000	190.19	211.11	274.44
Collision	$100 deductible	$466.53	$517.84	$574.70
	250 deductible	324.03	358.24	461.81
	500 deductible	261.95	290.77	378.01
Comprehensive	$50 deductible	$125.32	$137.85	$179.21
	100 deductible	93.99	104.33	135.62

Figure 7.1
Sample annual car insurance premiums.

Solution: Find the premium for $25/50,000 bodily injury and $25,000 property damage for a vehicle used to drive to work. Find the collision and the comprehensive premiums for a $100 deductible and a vehicle used to drive to work. Add all of the premiums together to find the total annual premium.

Bodily injury, $25/50,000 premium = $22.84

Property damage, $25,000 premium = $150.74

Collision, $100 deductible premium = $517.84

Comprehensive, $100 deductive premium = $104.33

Total annual premium = $22.84 + $150.74 + $517.84 + $104.33 = $795.75

Jenny's annual premium is $795.75.

PRACTICE PROBLEMS

7.26 Sabrina purchases 50/100/100 liability insurance for her auto. She uses her vehicle for pleasure only. What is her annual premium?

7.27 How much more will Sabrina's insurance cost if she adds collision and comprehensive coverage with a $250 deductible on collision and a $100 deductible on comprehensive?

Chapter

8

Insurance and Investments

Insurance and investments are vehicles to make your money grow, to protect your assets, and to provide for those who depend on you for financial support.

8.1 Life Insurance

Life insurance financially protects those who depend on you. Life insurance is a contract that provides for the *face value* of the policy to be paid to a *beneficiary*, or beneficiaries, after your death. Like other insurance, you will pay a monthly or annual *premium* for the insurance.

There are two basic types of life insurance:

• *Term life insurance* provides a death benefit for a certain number of years. If you live past the specified time period, the policy expires and no benefit is paid. After the policy expires, you must purchase a new policy in order to retain coverage. Term life insurance is the least expensive kind of insurance.

• *Permanent life insurance* insures you for your entire life. Part of the premium goes toward a cash value account that grows over time. Depending on what kind of permanent life insurance you have, you may be able to withdraw a portion of the cash value, take a loan against the cash value, or surrender the policy and take the cash value. Examples of permanent insurance include whole life, universal life, and variable universal life. Permanent life insurance is more expensive than term life insurance.

Premiums are calculated using variety of factors. Your age, gender, overall health and whether you are a smoker or not play significant roles in determining your premium. Table 8.1 shows sample premiums per $1,000 of life insurance for non-smokers in good health.

Table 8.1 Annual Premium per $1,000 of Life Insurance

Age of Insured	10-Year Term		Whole Life	
	Men	*Women*	*Men*	*Women*
20	$1.13	$1.06	$ 9.93	$ 8.90
25	$1.15	$1.09	$11.65	$10.60
30	$1.17	$1.12	$14.10	$12.81
35	$1.31	$1.25	$17.45	$15.90
40	$1.54	$1.48	$22.62	$20.55
45	$1.99	$1.90	$27.78	$25.24

To find the annual premium for a policy, divide the face value by $1,000 and multiply the result by the cost per $1,000 from the table.

$$\text{Premium} = \frac{\text{Face Value of Policy}}{\$1,000} \times \text{Cost per } \$1,000$$

Example: Sharon wants to purchase a $100,000 whole life policy. She is 30 years old, in good health, and a non-smoker. What is her monthly life insurance premium using Table 8.1?

Solution: Find the cost per $1,000 in the table. Divide the face value by $1,000, and then multiply by the cost per $1,000. Divide the annual premium by 12 to find the monthly premium.

30-year-old, whole life, female = $12.81.

$$\frac{\$100,000}{\$1,000} = 100$$

Annual premium = $100 \times \$12.81 = \$1,281$

Monthly premium = $\$1,281 \div 12 = \$106.75$

Sharon's monthly premium is $106.75.

PRACTICE PROBLEMS

8.1 Shawn is 40 years old, in good health, and a non-smoker. He wants to purchase a 10-year term life insurance policy with a face value of $500,000. What is the annual premium?

8.2 How much more will Shawn pay annually if he purchases $500,000 of whole life insurance instead of the 10-year term insurance?

Life Insurance Cash Value

Permanent life insurance policies build up a cash value after the first few years. The *cash value* is the amount of money the policy is worth if you cancel the policy. You may also be able to take out a loan up to the amount of the cash value. If you do not pay back the loan, the death benefit paid to the beneficiaries will be reduced by the amount owed for the loan.

The cash value earning rate on a whole life policy is guaranteed and could be represented by the figures in Table 8.2.

Table 8.2 Cash Value Table

Year of Policy	Cash/Loan Values per $1,000
1	$ 0
5	$ 10
10	$ 42
15	$ 80
20	$124
25	$174

Example: Mike has paid the premiums on a $500,000 whole life policy for 20 years. Using Table 8.2, how much can he borrow against the policy?

Solution: Find the factor in the table for a policy paid for 20 years. Divide the face value of the policy by $1,000. Multiply the result by the factor from the table.

Amount from the table = $124

Number of $1,000 units = $500,000 ÷ $1,000 = 500

Cash value = 500 × $124 = $62,000

Mike can borrow up to $62,000.

PRACTICE PROBLEMS

8.3 What is the cash value of a $100,000 whole life policy that has been paid for 25 years?

8.4 How much can you borrow against a $1,000,000 whole life policy in the 10th year of the policy?

8.2 Health Insurance

Like other insurance, *health insurance* protects against financial loss. Specifically, health insurance protects against financial loss due to medical bills. Some employers offer *group health insurance* as an employee benefit. Usually, the employer pays part or the entire premium, while the employee pays any remaining portion. If your employer does not offer group health insurance, you can purchase *private health insurance*, typically at a higher cost.

Example: Leona's employer provides group health insurance as an employee benefit. Her employer pays 25% of the $300 monthly premium that covers Leona and her family. How much does Leona pay in a year for health insurance?

Solution: Multiply the monthly premium by 25% to find out how much the employer pays. Subtract that amount from $300 to find out how much Leona pays per month. Multiply the monthly premium by 12 to find the cost for the year.

Employer's monthly portion $= \$300 \times 25\% = \$300 \times 0.25 = \$75$

Employee's monthly portion $= \$300 - \$75 = \$225$

Annual health insurance payment $= \$225 \times 12 = \$2,700$

Leona pays $2,700 annually for health insurance.

PRACTICE PROBLEMS

8.5 Jack has group health insurance through his employer. His employer pays 60% of the $500 monthly premium. What is Jack's annual health insurance cost?

8.6 Candice's employer does not provide health insurance. She obtains private health insurance for $428 per month. What is her annual premium?

Health Insurance Benefits and Costs

Most health insurance plans have an annual *deductible*, or the amount of expenses you must pay before the insurance company will begin to pay benefits. Once you have met your deductible, usually your insurance will pay a portion of your medicals bills, and you are responsible for the remaining portion. The payments that you make are called *co-insurance* or *co-payments*. For example, you may have to pay as co-insurance 20% of a medical bill, or you may have to make a co-payment of $25 per doctor's visit.

An insurance company will evaluate your medical bills before paying a medical claim. The services provided must be covered by the plan. In addition, the cost charged for the medical services must meet the insurance company's criteria for *customary and reasonable charges*. If the charge for services is more than the insurance company allows, then the insurance company will only pay benefits on the *maximum allowed charge*. If the insurance company has a contract with the provider, the excess charges may be written off, and you will not have to pay them. If the charges are not written off, you are responsible for your share of the allowable charges plus the all of the charges in excess of the allowed charge.

Example: Shane went on a ski trip and broke his leg. His medical expenses totaled $4,500. Shane has $300 of his deductible remaining and a co-insurance rate of 20%. The insurance company allowed $4,000 of the charges. Find the amount of the bill that Shane must pay if the excess charges are written off and the amount of the bill that Shane must pay if he is responsible for the excess charges.

Solution: Subtract the allowable charge from the actual charges to find the amount of excess charges. Subtract the remaining deductible from the allowed charges and multiply by the co-insurance rate to find the co-insurance amount. Add the deductible and the co-insurance amount to find out how much Shane owes if the excess charges are written off. To this amount, add the excess charges to find out how much Shane owes if he is responsible for the excess charges.

Excess charges = $4,500 − $4,000 = $500

Co-insurance amount = ($4,000 − $300) × 20% = $3,700 × 0.20 = $740

Amount without the excess charges = $300 + $740 = $1,040

Amount with the excess charges = $1,040 + $500 = $1,540

If Shane is not responsible for the excess charges, he must pay $1,040 of the medical expenses. If he is responsible for the excess charges, he must pay $1,540 of the medical expenses.

PRACTICE PROBLEMS

8.7 Last year, Nina had medical claims totaling $5,000. Her yearly deductible is $2,000. The insurance company has allowed all of the medical charges. If she has a co-insurance rate of 15%, how much of the $5,000 expenses will she pay?

8.8 Kirby pays $300 per month for health insurance. Last year his medical bills totaled $10,000. He has a $2,000 deductible and a co-insurance rate of 20%. The insurance company has allowed $9,000 of his medical bills, and Kirby is responsible for the excess charges. How much in total will Kirby pay for the year in medical insurance and expenses?

8.3 Disability Insurance

Disability insurance replaces a portion of your income if you become disabled and cannot work. Some employers provide group disability insurance coverage as an employee benefit. A typical group disability insurance policy will replace 60% of your salary for a maximum number of years or until your retirement.

Example: Nathan becomes disabled and cannot work. His disability insurance policy pays 60% of his average annual salary for the last three years. The last three years, Nathan earned annual salaries of $57,000, $65,000, $70,000. What is Nathan's monthly disability payment?

Solution: Find the average of the three salaries by adding them together and dividing by 3. Multiply the average by 60% to find the annual disability payment. Divide the annual disability payment by 12 to find the monthly payment.

$$\text{Average salary} = \frac{\$57,000 + \$65,000 + \$70,000}{3} = \frac{\$192,000}{3} = \$64,000$$

Annual disability payment = $64,000 × 60% = $64,000 × 0.60 = $38,400

Monthly disability payment = $38,400 ÷ 12 = $3,200

Nathan's monthly disability payment is $3,200.

PRACTICE PROBLEMS

8.9   How much would Nathan's monthly disability payment be if the payments are based on his last year's salary instead of the average of the last three years?

8.10   If Nathan receives the $3,200 per month, what percentage of his current salary is he receiving?

8.4 Bonds

When an incorporated business or the government needs to raise money, it may sell bonds. A *bond* is an agreement to repay the *face value* or the *par value* of the bond, usually $1,000, on a certain date. In addition to the payment at the maturity date, the bondholder receives a specific rate of interest paid semiannually.

Bonds can be bought and sold before the maturity date of the bond. The *market value* is the current selling price of a bond and may be higher or lower than the par value of the bond. The selling price is quoted as a percentage of the bond's par value. A selling price of 98.25 means that the bond is selling for 98.25% of its par value.

Example: Find the price of a single $1,000 bond if the bonds are selling at 97.33.

Solution: Convert the market price to a decimal and multiply by the par value of the bond

Market price = 97.33% = 0.9733

Price = $1,000 × 0.9733 = $973.30

The bond is selling for $973.30.

PRACTICE PROBLEMS

8.11 Inglewood School District is selling $1,000 bonds to raise money to build a new school. If the market price is 101.275, what is the price of one bond?

8.12 The market price of $1,000 bonds issued by the city of Binkleberg is 74.875. What is the price of 10 bonds?

Bond Interest

Bonds typically pay interest semiannually at the interest rate stated on the bond. The interest is based on the bond's par value. To find the interest on a bond, you can use the simple interest formula.

Interest = Principal × Rate × Time, where the principal is the par value.

$$I = P \times R \times T$$

Example: Martha owns 10 bonds with a par value of $1,000. The bonds pay interest semiannually at a rate of 5.5%. How much will Martha receive semiannually?

Solution: To find the interest on a single bond, multiply the par value by the interest rate expressed as a decimal and by the time, half of a year.

Interest rate = 5.5% = 0.055

Interest on one bond = $1,000 × 0.055 × $\dfrac{1}{2}$ = $27.50

Interest on 10 bonds = $27.50 × 10 = $275

Martha will earn $275 in interest semiannually.

PRACTICE PROBLEMS

8.13 Julio owns 10 bonds with a par value of $1,000. The bonds pay 8.4% interest semiannually. How much interest will Julio receive semiannually?

8.14 Greg owns five bonds with a par value of $1,000. The bonds pay 7.62% interest semiannually. How much interest will Greg receive annually?

Current Yield

The *current yield* expresses the rate of return on a bond if you held it until maturity, based on the price you paid for the bond.

If the bond price is lower than its par value, the current yield will be higher than the bond's interest rate. If the bond price is higher than its par value, the current yield will be lower than the bond's interest rate.

To find the current yield, divide the annual interest by the bond price. Current yield is usually expressed as a percent, so convert the decimal to a percent by multiplying by 100 or moving the decimal point two places to the right.

$$\text{Current Yield} = \frac{\text{Annual Interest}}{\text{Bond Price}}$$

Example: What is the current yield on a $1,000 bond with an 8% interest rate priced at 78.375? Round to the nearest tenth of a percent.

Solution: Calculate the annual interest on the bond by multiplying the par value by the interest rate. Find the price of the bond by multiplying the par value by the market price. Divide the annual interest by the bond price and change the result to a percent.

Interest = $1,000 × 8% = $1,000 × 0.08 = $80

Market price = 78.375% = 0.78375

Bond price = $1,000 × 0.78375 = $783.75

$$\text{Current yield} = \frac{\$80}{\$783.75} = 0.102 = 10.2\%$$

The current yield on the bond is 10.2%.

PRACTICE PROBLEMS

8.15 The Klinket Company has $1,000 6% bonds that are offered at 102.458. What is the current yield to the nearest tenth of a percent?

8.16 The semiannual interest on $1,000 Lansing Manufacturing bonds is $53.25. If the current price is 93.275, what is the current yield to the nearest tenth of a percent?

Cost of Buying and Selling Bonds

Bonds are usually purchased through a broker who will receive a commission for the bonds that are bought and sold on your behalf.

In addition, when a bond is sold between interest payment dates, the new owner of the bond receives all of the interest at the next interest payment. However, as a part of the purchase, the new owner must pay the previous owner the interest that has accrued since the last interest payment.

To find the total cost to purchase bonds, add the price of the bonds, the commission and the accrued interest that must be paid to the previous owner.

Example: Myra purchases 10 bonds with a par value of $1,000 at 102.75. If she must pay accrued interest of $25 per bond and a commission of $5 per bond, what is the total cost of the transaction?

Solution: Find the cost of one bond by multiplying the decimal value of the price by $1,000. Add the accrued interest and commission on a single bond to find the total cost per bond. Multiply by 10, the number of bonds purchased.

Market price = 102.75% = 1.0275

Price for one bond = $1,000 × 1.0275 = $1,027.50

Total cost for one bond = $1,027.50 + $25 + $5 = $1,057.50

Total cost for 10 bonds = $1,057.50 × 10 = $10,575

Myra will spend $10,575 to purchase the bonds.

When you sell bonds, you may be paid accrued interest by the new owner. You will have to pay a commission on the sale of the bond. To find the amount received from the sale of bonds, add the price of the bonds and the accrued interest, and then subtract the commission.

Example: From the previous example, Myra sold the 10 bonds when the market price was 108.295. To sell the bonds, she pays the same $5 commission per bond. She will receive $58 per bond in accrued interest at the sale. How much will Myra receive when she sells the bonds?

Solution: Find the selling price of one bond by multiplying the decimal value of the market price by the par value of the bond. Add the accrued interest. Subtract the commission to find the amount Myra receives for one bond. Multiply the result by 10 to find the amount she will receive for 10 bonds.

Market price = 108.295% = 1.08295

Price of one bond = $1,000 × 1.08295 = $1,082.95

Total received for one bond = $1,082.95 + $58 − $5 = $1,135.95

Total received for 10 bonds = $1,135.95 × 10 = $11,359.50

Myra receives $11,359.50 on the sale of the bonds.

PRACTICE PROBLEMS

8.17 Find the total investment in five bonds with a par value of $1,000 that are selling for a market price of 112.225, a broker's commission of $4 per bond, and accrued interest due of $33 per bond.

8.18 Find the amount received on the sale of 10 bonds with a par value of $1,000 and a market price of 88.253 if the commission is $6 per bond and the accrued interest is $125 per bond.

8.5 Stocks

When a company incorporates it can issue *stock*. When you purchase stock in a company, you are a *shareholder*, and you own a part of the company. When a company earns a profit, the profit may be wholly or partially reinvested in the company or paid to the shareholders as *dividends*.

A corporation that issues stock sets the price, or the *par value*, of the stock. As the stock is bought and sold on the stock market, its *market price* will change based on many factors. You can find daily stock reports in newspapers or online.

You purchase stock through a stockbroker, who earns a commission when you buy or sell stock. The commission rate paid to a stockbroker can vary greatly, based on the amount of service the broker provides. To find the total cost of a stock transaction, add the price of the stock and the commission.

Example: Niam purchases 100 shares of stock in Nixta Corporation at $35. The broker charged a 1% commission. Find the total cost of the stock.

Solution: Multiply the number of shares by the cost per share to find the cost of the stocks. Multiply the cost of the stocks by 1% to find the commission. Add the cost of the stocks and the commission.

Cost of the stocks = 100 × $35 = $3,500

Commission = $3,500 × 1% = $3,500 × 0.01 = $35

Total cost = $3,500 + $35 = $3,535

The total cost to Niam is $3,535.

PRACTICE PROBLEMS

8.19 Alexis purchased 50 shares of stock at $14.80 per share. The broker charged a $5 commission. Find the total cost of the stock.

8.20 Jim purchased 150 shares of stock at $32.95 per share. The broker charged a 4% fee. What was the total cost of the stock?

Stock Dividends

As part owners of the company, stockholders can be paid a share of the profits, usually paid quarterly as dividends. There is no guarantee of dividends for stockholders since there may be no profit, or the board may decide to re-invest the profits in the company. Dividends may be expressed as a dollar amount, or as a percent of the par value of the stock.

Example: Bailey owns 200 shares of stock in Juxta Energy Corporation. The stock has a par value of $100 per share. The company declares a dividend of 3.7% for the fourth quarter. How much should Bailey receive in dividends for the fourth quarter?

Solution: Find the dividend on a single share of stock by multiplying the par value by the dividend percent. Multiply the dividend on one share of stock by the number of shares owned.

Dividend on one share $= \$100 \times 3.7\% = \$100 \times 0.037 = \$3.70$

Dividend on 200 shares $- \$3.70 \times 200 = \$740$

Bailey's fourth quarter dividend is $740.

Practice Problems

9.21 Kim owns 50 shares of stock in Maxwell Manufacturing. The company announces a dividend of $2.38 per share. How much should Kim receive in dividends?

8.22 Sandra owns 200 shares of stock with a par value of $100 per share. If the current quarter's dividend is 1.8%, how much should Sandra receive?

Cost Yield

The *cost yield* of a stock, or the rate of income received for your investment, can be found by dividing the annual dividends by the amount invested in the stocks including any expenses or commission paid. The cost yield is typically expressed as a percent, so change the decimal to a percent by multiplying by 100 or moving the decimal point two places to the right.

$$\text{Cost Yield} = \frac{\text{Annual Dividends}}{\text{Total Cost of Stock}}$$

Example: Josiah purchased 20 shares of Marquet Industries stock at $30 per share. The broker's commission was $5. If the stock pays an annual dividend of $2.25 per share, what is the cost yield for this investment, rounded to the nearest tenth of a percent?

Solution: Multiply the price by the number of shares to find the cost of the stock. Add the commission to find the total cost. Multiply the annual dividend by the number of shares of stock to find the total annual dividends. Divide the total annual dividends by the total cost of the stock, and then change to a percent.

Cost of the stock = $30 \times 20 = $600

Total cost of the stock = $600 + $5 = $605

Total annual dividends = $2.25 \times 20 = $45

$$\text{Cost yield} = \frac{\$45}{\$605} = 0.074 = 7.4\%$$

Practice Problems

8.23 What is the cost yield on stock that cost $4,850 and pays annual dividends of $336?

8.24 Zachariah purchased 30 shares of $100 par value stock for $22 per share and a 2% commission. The stock paid an annual dividend of 3.9%. What is the cost yield for this investment, rounded to the nearest tenth of a percent?

Selling Stock

When you sell stock, you will pay a commission and other fees. The net proceeds of a stock sale is the market price minus the commission and other fees.

Net Proceeds = Market Price − (Commission + Other Fees)

Example: Shondra sells 100 shares of Telenet Communications stock for $18 per share. She pays a commission and other fees of $56. What are her net proceeds from the sale?

Solution: Multiply the number of shares by the market price and subtract the commission and fees.

Net proceeds $= 100 \times \$18 - \$56 = \$1{,}744$

Shondra's net proceeds are $1,744.

One way to evaluate your stock investment is to calculate the *return on investment (ROI)*. First calculate the total gain on your investment by adding the proceeds of a stock sale to the dividends you received while you owned the stock and subtracting the original cost, including the cost of the stock and the commission. Divide the total gain by the original cost of the stock, and then change the decimal to a percent.

- Original Cost = Stock Price + Commission
- Total Gain = Net Proceeds + Dividends − Original Cost

- Return on Investment (ROI) $= \dfrac{\text{Total Gain}}{\text{Original Cost}}$

Notice that the return on investment calculation does not factor in time. For example, you could calculate the return on investment for two different investments and find that the return on investment was 10% on both. However, one investment may make a 10% return on investment in one year, while the other investment may take 20 years to make a 10% return on investment.

Example: Jason purchased 100 shares of stock at $12 per share, and paid a commission of $10. He sold the stock at $15 per share and paid a commission and fees totaling $25. While he owned the stock, he earned $123 in dividends. What is Jason's return on investment to the nearest tenth of a percent?

Solution: Calculate the original cost of the stock by multiplying the number of shares by the purchase price and adding the commission. Find the net proceeds by multiplying the number of shares by the selling price and subtracting the commission and fees. Find the total gain by adding the net proceeds and the dividends and subtracting the original cost. Divide the total gain by the original cost and change the decimal to a percent.

Original cost = $100 \times \$12 + \$10 = \$1,210$

Net proceeds = $100 \times \$15 - \$25 = \$1,475$

Total gain = $\$1,475 + \$123 - \$1,210 = \$388$

Return on investment = $\dfrac{\$388}{\$1,210} = 0.321 = 32.1\%$

Jason's return on investment is 32.1%.

PRACTICE PROBLEMS

8.25 What are the net proceeds of selling 200 shares of stock at $9.45 with a $25 commission?

8.26 Nolan purchased 50 shares of stock at $56 per share and paid a commission of 1%. He sold the stock at $66 per share and paid a commission and fees totaling $35. While he owned the stock, he earned $60 in dividends. What is Nolan's return on investment to the nearest tenth of a percent?

8.6 Mutual Funds

A *mutual fund* is a collection of stocks, bonds, and other assets held by a mutual fund company. Investors in the fund own shares of the fund, which represent a portion of the company's holdings.

Income is generated on the dividends of the assets and passed on to the investors as *distributions*. Investors also receive distributions when the fund sells stocks and bonds at a profit. Investors can also choose to reinvest distributions in the fund to increase the number of shares held.

Mutual fund shares are bought and sold based on the fund's *net asset value*, or *NAV*. The net asset value is the total value of the fund's investments, less any debt, divided by the number of shares held by investors. The NAV of a mutual fund can change daily, and even multiple times per day.

All mutual fund companies charge an ongoing yearly fee.

Some companies also charge a commission, called a *load*. A *no-load fund* has no commission, while a *load fund* charges a commission.

The price of a share in a no-load mutual fund is the NAV of the fund. The price of a share in a load fund, called the *public offering price*, is the NAV plus the commission. Typically an investor will invest a dollar amount in a mutual fund rather than buying a certain number of shares. It is possible to own fractional shares in a mutual fund company. Statements will usually show the number of shares rounded to the nearest thousandth.

Example: Jack invests $1,000 in a no-load mutual fund with a NAV of $15.22. How many shares of the mutual fund does he own?

Solution: Divide the amount invested by the NAV. Round to the nearest thousandth.

Shares purchased = $1,000 ÷ $15.22 = 65.703

Jack owns 65.703 shares in the mutual fund.

Example: Margie invests $1,000 in a mutual fund with a NAV of $7.39 and a public offering price of $8.13. How many shares of the mutual fund does she own?

Solution: Divide the amount invested by the public offering price. Round to the nearest thousandth.

Shares purchased = $1,000 ÷ $8.13 = 123.001

Margie owns 123.001 shares in the mutual fund.

PRACTICE PROBLEMS

8.27 If you invest $2,500 in a no-load mutual fund with a NAV of $9.82, how many shares will you own? Round to the nearest thousandth.

8.28 If you invest $2,500 in load fund with a NAV of $9.82 and a public offering price of $10.75, how many shares will you own? Round to the nearest thousandth.

Mutual Fund Commission

To find the amount of commission on a load fund, subtract the NAV from the public offering price. The commission or load may be reported by the mutual fund company as a dollar or percent amount.

Commission (Load) = Public Offering Price − Net Asset Value

Example: What is the commission on a load fund that has a NAV of $12.83 and a public offering price of $13.25?

Solution: Subtract the NAV from the public offering price to find the commission.

Commission = $13.25 − $12.83 = $0.42

The commission on the load fund is $0.42 per share.

PRACTICE PROBLEMS

8.29 How much commission is being charged on a mutual fund with a public offering price of $6.75 and a NAV of $6.03?

8.30 How much commission is being charged on a mutual fund with a public offering price of $18.22 and a NAV of $17.29?

Profit or Loss

When you *redeem*, or sell, your shares back to the mutual fund company, you are paid the NAV for each share or part of a share that you redeem. The proceeds can be calculated by multiplying the shares redeemed by the net asset value.

Proceeds = Number of Shares Redeemed × Net Asset Value

To calculate the profit or loss, you need to account for the distributions you have received and fees that you have paid. Often investors have distributions reinvested in the fund to buy more shares and the fees taken from the fund by redeeming shares. In this case, you can calculate your profit or loss in a mutual fund by subtracting the amount invested from the proceeds. If the result is positive, there is a profit. If the result is negative, there is a loss.

Profit or Loss = Proceeds − Amount Invested

Example: Sara invested $10,000 in a mutual fund with a public offering price of $5.97. Distributions were reinvested and fees were paid through the redemption of shares. After five years, she owns 1,845.397 shares and redeems them at a NAV of $7.38. What is her profit or loss?

Solution: Multiply the number of shares owned by the current NAV to find the proceeds from the redemption of shares. Subtract $10,000 from the proceeds to determine her profit or loss.

Proceeds = 1,845.397 × $7.38 = $13,619.03

Profit or Loss = $13,619.03 − $10,000 = $3,619.03

Since $3,619.03 is positive, there is a profit.

Sara earned a profit of $3,619.03.

PRACTICE PROBLEMS

8.31 What are the proceeds when you redeem 10,487.33 shares of a mutual fund with a NAV of $10.36?

8.32 Greg invested $6,000 in a mutual fund. He reinvested the dividends and paid the fees with the redemption of shares. After 10 years, he owns 867.508 shares and redeems them at a NAV of $6.34. What is his profit or loss?

8.7 Retirement Investments

When you retire, you may receive retirement income from several different sources, such as Social Security benefits, a pension plan through your previous employer, and income from investments.

Pension Income

A *pension plan* is a retirement plan where contributions are made to a pool of funds that are invested for the employee's future benefit. In a *defined contribution plan*, you and your employer both contribute to your pension fund. The money you have available at retirement depends on how well the investments in the fund performed.

In a *defined benefit plan*, you are guaranteed an amount per month when you retire. As an employee, you may or may not be required to contribute money to the plan. The amount you receive will depend on your length of service and your wages.

Qualified pension plans are *tax-deferred* investments because you do not pay taxes on the money until it is withdrawn after retirement.

Example: Rosa participates in a defined benefit plan through her employer. She retires after 25 years of employment and will receive 2.5% of the average of her last three years of salary for each year that she was employed. If her last three years of salary are $73,000, $84,000 and $86,000, what is her monthly pension?

Solution: Multiply the pension rate by the number of years of service to find the total pension rate. Find the average of her last three years of salary by adding the salaries and dividing by 3. Multiply the total pension rate by the average salary to find the annual pension. Divide the annual pension by 12 to find the monthly pension.

Total pension rate $= 2.5\% \times 25 = 62.5\%$

$$\text{Average salary} = \frac{\$73,000 + \$84,000 + \$86,000}{3} = \frac{\$243,000}{3} = \$81,000$$

Annual pension $= \$81,000 \times 62.5\% = \$81,000 \times 0.625 = \$50,625$

Monthly pension $= \$50,625 \div 12 = \$4,218.75$

Rosa's monthly pension is $4,218.75.

PRACTICE PROBLEMS

8.33 How much would Rosa have received per month if her pension was based on her last year's salary instead of the average of the last three years?

8.34 How much more would Rosa have received per month if she worked three more years and her average salary for those three years was $92,000?

Required Minimum Distributions

Other retirement investments may be tax free or tax deferred. One type of retirement account is called an *individual retirement account*, or *IRA*. The government determines how much money may be invested yearly in an IRA. Contributions to a *traditional IRA* are tax-deductible in the year you make the contribution, and the money grows in the account tax deferred. Contributions to a *Roth IRA* are not tax-deductible, but since you paid taxes on the money going into the account, you do not pay taxes on any of the money when you withdraw it for retirement. The money in a Roth IRA grows tax free.

Employee sponsored defined contribution plans and traditional IRAs have *required minimum distributions*. Once you reach 70½ years of age, you are required to begin withdrawing some of the money. The required distribution is calculated using divisors that change based on your age. Divisors for different ages are shown in Table 8.3.

Table 8.3 Required Minimum Distribution Divisors

Age	Divisor	Age	Divisor
70	27.4	78	20.3
71	26.5	79	19.5
72	25.6	80	18.7
73	24.7	81	17.9
74	23.8	82	17.1
75	22.9	83	16.3
76	22.0	84	15.5
77	21.2	85	14.8

To calculate the minimum distribution, divide the value of the account by the divisor that corresponds to the age of the retiree.

Example: Jonas is 75 years old and owns a traditional IRA worth $350,000. What is his required minimum distribution for the year?

Solution: Find the divisor that corresponds to an age of 75. Divide the value of the IRA by the divisor to find the minimum distribution for the year.

Divisor for 75 years old = 22.9

Required minimum distribution = $350,000 ÷ 22.9 = $15,283.84

Jonas is required to withdraw $15,283.84 from his traditional IRA this year.

PRACTICE PROBLEMS

8.35 The following year, Jonas's traditional IRA has a value of $338,000. What is his required minimum distribution?

8.36 William's employer-sponsored retirement program has a value of $200,000 when William is 82 years old. What is the required minimum distribution for the year?

Chapter

9

Budgets

With the ease of obtaining credit and so many ways and opportunities to spend money, many people find themselves in financial trouble. Even if you aren't in financial trouble, understanding where your money goes is an important step in managing your money and avoiding trouble. A *budget* is a spending plan that can help you manage your money and reach financial goals.

9.1 Average Monthly Expenses

Understanding where your money is currently going is the first step in making a budget. A simple activity is to track all of your expenses by writing down all the money you spend for a certain length of time.

Example: Cindy tracks her expenses for a single week and discovers that she spent $24 at the local coffee shop. At that rate, how much will she spend in one year at the coffee shop?

Solution: Multiply the amount spent per week by 52, the number of weeks in a year.

Amount spent per year = $24 \times 52 = \$1,248$

At the current rate, Cindy will spend $1,248 per year at the coffee shop.

Once you have tracked expenses for several months, you can calculate your average monthly expenses.

$$\text{Average Monthly Expenses} = \frac{\text{Total Expenses}}{\text{Number of Months}}$$

Example: LaToya tracked her family's expenses for three months and found that she spent the following amounts on food each month: $482.50, $508.16, and $522.37. What is her family's average monthly expense for food, rounded to the nearest dollar?

Solution: Find the sum of the expenses and then divide by 3, the number of months.

$$\text{Average monthly expense} = \frac{\$482.50 + \$508.16 + \$522.37}{3} = \frac{\$1,513.03}{3} = \$504.34$$

LaToya's family spends about $504 per month on food.

Certain expenses, such as insurance premiums, property taxes, and car registration fees, may be paid quarterly, semi-annually, or yearly, and may not make it on a list of expenses tracked for just a few months. These expenses should be averaged over the payment period so you can save the money to be paid when it is due.

Example: Shawn's average transportation expense excluding insurance is $325 a month. His annual auto insurance premium is $1,236. How much should Shawn save per month to have enough to pay the annual premium? What is the total monthly transportation expense including insurance?

Solution: Divide the annual auto insurance premium by 12 to find the amount needed to save each month. Add the monthly premium amount to the monthly auto expense to find the total monthly auto expense.

Monthly amount needed for insurance = $1,236 ÷ 12 = $103

Total monthly auto expense = $325 + $103 = $428

Shawn needs to save $103 per month to pay his annual auto insurance premium. His monthly transportation expense including insurance is $428.

PRACTICE PROBLEMS

9.1 Jack likes to go out to lunch during the workweek. If he spends an average of $250 per month going out to lunch, how much will he spend in a year?

9.2 Bobby spent the following amounts on clothing over the last three months: $28.45, $134.27, $75.63. What is his average monthly expenditure on clothing?

9.3 Yvette's monthly housing expense excluding property taxes is $955. If she estimates her annual property tax bill to be $1,500, what is her total monthly housing expense including property taxes?

9.2 Creating and Adjusting a Budget

Once you have an idea of where your money is going, you should evaluate your spending habits. Among other things, you should consider if you are spending more than you make, if you are overspending in certain categories, and if you are saving money at a rate that will help you meet your financial goals.

Many financial advisors suggest the following standards by which to evaluate spending:

- Save 10–20% of your income in a savings account or other investments.
- Spend no more than 25–30% of your net pay on housing.
- Spend no more than 20% of your net pay on transportation.

Notice that the guidelines for savings is based on your income or gross pay, while housing and transportation are based on your net or take-home pay.

Example: Ira earns a salary of $35,000. After taxes, his net income is $2,350 per month. If he saves $200 per month and spends $800 per month on housing and $600 per month on transportation, evaluate his spending habits.

Solution: Multiply the amount he saves per month by 12 and then divide by his annual salary to determine the percent of his income he is saving. Divide his monthly housing and transportation expenses by his net monthly income to determine the percent of his net pay he is spending in each of these areas. Evaluate by comparing to the preceding guidelines.

Amount saved per year = $200 × 12 = $2,400

Percent of income saved = $2,400 ÷ $35,000 = 0.07 = 7%

Percent of net pay spent on housing = $\dfrac{\$800}{\$2,350}$ = 0.34 = 34%

Percent of net pay spent on transportation = $\dfrac{\$600}{\$2,350}$ = 0.26 = 26%

Using the given guidelines, Ira is saving too little and spending too much on housing and transportation.

A budget should reflect the spending averages that you found when you tracked your expenses as well as incorporate the changes you may want to make in your spending habits. A budget should be broken into manageable spending categories.

Example: In the previous example, Ira discovered that he wasn't saving enough money. His goal for this year is to save 20% of his income. How much should he save per month?

Solution: Multiply his annual salary by 20% and divide the result by 12.

Annual savings budget = $35,000 × 20% = $35,000 × 0.20 = $7,000

Monthly savings budget = $7,000 ÷ 12 = $583.33

Ira should budget $583 per month for savings to save 20% of his income.

PRACTICE PROBLEMS

9.4 Bella's income is $45,000 per year, and her monthly take home pay is $3,180 per month. Bella has been saving $750 per month between a savings account and investments. She spends $1,050 per month on housing and $500 per month on transportation. Evaluate her savings and spending habits.

9.5 Bella's salary is increased to $55,000 per year. If she wants to save 20% of her income, how much should she save per month?

A good budget will address not only current spending habits and desired changes, but also estimated future income and expenses. For example, you may be expecting a raise, or you may have a child that will begin driving and need car insurance.

A budget can cover any desired time period, although most budgets are monthly, quarterly, or yearly. A budget that uses percents rather than dollar amounts can be used for multiple time periods.

Example: The Simpson family has tracked their spending and identified areas where they want to change their spending habits. They have accounted for any expected changes in expenses. Table 9.1 shows the percent of take-home pay they have allotted for each category.

Table 9.1 Simpson Family Budget

Category	Percent of Net Income
Housing	25%
Utilities	5%
Life Insurance	5%
Transportation	10%
Food	15%
Clothing	2%
Entertainment	3%
Savings	15%
Charitable Giving	10%
Health Care	10%

If the Simpson's monthly take-home pay is $4,000, how much have the Simpson's budgeted for housing and transportation each month? How much have they budgeted yearly for food?

Solution: Multiply the monthly take-home pay by the percent budgeted for housing and transportation. To find the amount budgeted yearly for food, multiply the monthly take home pay by 12 to find the yearly take-home pay, and then multiply by the percent budgeted for food.

Housing budget = $4,000 × 25% = $4,000 × 0.25 = $1,000 per month

Transportation budget = $4,000 × 10% = $4,000 × 0.10 = $400 per month

Yearly take-home pay = $4,000 × 12 = $48,000

Food budget = $48,000 × 15% = $48,000 × 0.15 = $7,200 per year

The Simpsons have budgeted $1,000 per month for housing, $400 per month for transportation and $7,200 per year for food.

After you have created your budget and worked with it for a period of time, you may find that your actual spending does not match what you have allocated for each budget category. If your total spending is not over your budget, you may simply want to transfer money between budget categories.

Example: The Simpson's have calculated the six-month average of their spending to compare to their budget (see Table 9.2). Suggest how money could be transferred between budget categories to align their budget with their spending.

Table 9.2 Simpson Family Budget

Category	Percent Budgeted	Dollars Budgeted	6-Month Average
Housing	25%	$1,000	$1,000
Utilities	5%	$ 200	$ 175
Life Insurance	5%	$ 200	$ 200
Transportation	10%	$ 400	$ 375
Food	15%	$ 600	$ 650
Clothing	2%	$ 80	$ 80
Entertainment	3%	$ 120	$ 120
Savings	15%	$ 600	$ 600
Charitable Giving	10%	$ 400	$ 400
Health Care	10%	$ 400	$ 400
TOTAL		$4,000	$4,000

Solution: Identify the categories in which the Simpson's spent more or less than was budgeted. Change those categories to reflect spending.

Utilities = $200 − $175 = $25

The Simpson's spent $25 less per month on utilities.

Transportation = $400 − $375 = $25

The Simpson's spent $25 less per month on transportation.

Food = $650 − $600 = $50

The Simpsons spent $50 more per month on food.

Transfer $25 each from the utilities and transportation categories into the food category.

PRACTICE PROBLEMS

9.6   The Simpsons decide that they want to change their budget for savings to 20% of their take-home pay. What will they have to do in order to make that happen?

9.7   How much more money per year will be budgeted for savings if they change the savings budget to 20%?

More often than not, adjusting a budget requires reducing spending. If you are spending more money than you make each month or you are not saving enough money each month, you must reduce spending or increase your income. However, for many people, an increase in income leads to an increase in spending, and they continue to spend more than they make. Reducing spending is the most effective way to address budget issues.

There are many books, articles, and websites devoted to reducing spending. A few quick ways to reduce spending include

- Go out to eat less often.
- Get rid of stuff you don't use and the storage space you may be renting for it.
- Make your home more energy efficient.
- Entertain at home.
- Shop for the best buys on everyday items as well as large items.
- Evaluate personal expenses, such as cell phone, Internet, and expanded television service plans.

9.3 Best Buys

Shopping for the best buys on items can help you reduce spending in certain budget categories. You can compare the cost of two items of different size by calculating the *unit price.*

$$\text{Unit Price} = \frac{\text{Cost}}{\text{Number of Units}}$$

Example: Sparkling White mouthwash costs $2.48 for 16 oz. Fresh Breeze mouthwash costs $1.78 for 9 oz. Which brand of mouthwash costs less per ounce?

Solution: Find the unit cost of each mouthwash by dividing the cost by the number of ounces. Compare the unit costs and choose the lowest price.

$$\text{Sparkling White unit cost} = \frac{\$2.48}{16\text{ oz}} = \$0.155 \text{ per ounce}$$

$$\text{Fresh Breeze unit cost} = \frac{\$1.78}{9\text{ oz}} = \$0.198 \text{ per ounce}$$

Sparkling White costs almost 16 cents an ounce, while Fresh Breeze costs almost 20 cents per ounce. Sparkling White has the lowest unit cost.

Some stores will post a unit price for products. However, before using the posted unit price for comparison, make sure that the unit price is for the same unit of measure. For example, for one product, the posted unit price may be the cost per ounce, while for another product the posted price may be the cost per pound.

Sometimes you can save money by purchasing items in a larger quantity, but beware because the larger package is not always the best deal.

Example: You can purchase a package of four rolls of Clean-o paper towels for $6.45, or you can purchase a package of 24 rolls of Clean-o paper towels for $32. Do you save money by buying the larger package? If so, how much do you save?

Solution: Divide 24 by 4 to find out how many of the smaller packages you would have to purchase to have the same number of rolls as the larger package. Multiply the number of packages needed by the cost per package for four rolls. Subtract to find out how much you save.

Number of smaller packages needed = $24 \div 4 = 6$ packages

Cost for six smaller packages = $6.45 \times 6 = $38.70

24-roll package = $32

24-roll package is cheaper than six four-roll packages.

Savings = $38.70 − $32 = $6.70

You will save $6.70 if you purchase the larger package.

PRACTICE PROBLEMS

9.8 Tidalwave laundry detergent sells for $16.50 for 125 ounces. Fresh and Clean laundry detergent sells for $9.36 for 85 ounces. If you will use the same amount of laundry detergent per load, which laundry detergent is the best buy?

9.9 You can buy a package of four batteries for $5.89 or a package of 36 batteries for $55.25. Do you save money by buying the larger package? If so, how much do you save?

9.4 Optional Personal Expenses

Many young people have grown up with the luxury of cell phones, Internet service, and expanded television service and view these items and services as necessary. However, if you are in a position where you have to reduce your spending, these expenses, along with others, need to be evaluated. In some cases, you may find ways to reduce these expenses, rather than eliminate them.

Example: Ginger is evaluating her cell phone plan to see if she can save money. She currently has a plan that has unlimited calls and messaging for $70 per month. Taxes add 15% to her phone bill each month. On average, she uses about 420 minutes per month. Her carrier has a plan that offers 500 minutes per month for $45 per month. She estimates the taxes will be about 15% of the bill. How much can she save per month by switching plans? How much can she save in a year by switching plans?

Solution: Multiply the monthly rates by 15% to find the taxes. Add the taxes to the monthly rate to find out the total monthly expenses for each cell phone plan. Subtract the lower cell phone rate from the larger to find the monthly savings. Multiply the monthly savings by 12 to find the annual savings.

Taxes on current rate = $70 × 15% = $70 × 0.15 = $10.50

Current monthly expense = $70 + $10.50 = $80.50

Taxes on new rate = $45 × 15% = $40 × 0.15 = $6.75

New monthly expense = $45 + $6.75 = $51.75

Savings per month = $80.50 − $51.75 = $28.75

Savings per year = $28.75 × 12 = $345

Ginger can save $28.75 per month and $345 per year by switching cell phone plans.

PRACTICE PROBLEMS

9.10 From the previous example, Ginger's carrier also offers a pay-as-you-go phone for $0.08 per minute with free unlimited texting. If taxes remain at 15% of the bill, and she uses an average of 420 minutes per month, how much should she expect to pay per month?

9.11 Another phone company's pay-as-you-go plan charges $10 for 30 minutes of airtime and unlimited texting. If taxes are 15%, how much should Ginger expect to pay per month with this plan if she uses 420 minutes per month?

Internet connection expenses may include equipment and installation costs, monthly fees, and yearly subscriptions for Internet security and filtering software.

Example: The Thomlinson family plans to sign up for Internet service. The provider advertises a monthly subscription rate of $40 plus 15% in taxes. In addition, there is an installation fee of $100 and a one-time equipment fee of $50.

The Thomlinsons plan to invest in Internet security software that costs $80 per year, and Internet filtering software that costs $50 per year. What is the cost of Internet service for the first year? After the first year, what is the cost per month?

Solution: Multiply the monthly subscription rate by 15% to calculate the taxes. Add the taxes to the monthly subscription rate and multiply by 12 to get the yearly subscription rate. Add the installment and equipment fees and the software costs to find the total cost for the first year. Add the yearly subscription rate and the software costs to find the total yearly rate after the first year, and then divide by 12 to find the total monthly rate after the first year.

Tax on monthly subscription rate = $40 × 15% = $40 × 0.15 = $6

Total monthly subscription rate = $40 + $6 = $46

Yearly subscription rate = $46 × 12 = $552

Total first year cost = $552 + $100 + $50 + $80 + $50 = $832

Total annual cost after first year = $552 + $80 + $50 = $682

Total monthly cost after first year = $682 ÷ 12 = $56.83

The total cost for Internet service in the first year is $832. After the first year, the total cost per month is $56.83.

PRACTICE PROBLEM

9.12 The Internet service provider that Thomlinsons are planning to sign up with has a special. If you sign up for two years of service, they will wave the installation and equipment fees and reduce the monthly fee to $35 per month. What will be their monthly cost if they sign up for the special and purchase the Internet security and filtering software?

While local television service is free, expanded television programming provided by cable or satellite is a paid service.

Example: The Johanson family signs up for satellite television. During the first year, the programming portion of their bill will be $34.99. After the first year, the programming portion of their bill will increase to $63.99 per month. Other monthly fees include HD access $10, movies $26, sports $10, and DVR and HD $6. There is a 6% tax on the programming portion of the bill only. How much do they need to budget per month for satellite television for the first year? How much do they need to budget per month for satellite television after the first year?

Solution: Multiply the programming charge by 6% to find the taxes. Add the taxes to the programming cost and all the other fees to find the total bill. Repeat for the fees after the first year.

Programming tax, first year = $34.99 × 6% = $34.99 × 0.06 = $2.10

Monthly bill, first year = $34.99 + $2.10 + $10 + $26 + $10 + $6 = $89.09

Programming tax, after first year = $63.99 × 6% = $63.99 × 0.06 = $3.84

Monthly bill, after first year = $63.99 + $3.84 + $10 + $26 + $10 + $6 = $119.83

The Johanson family must budget $89.09 per month for the first year for satellite service and $119.83 per month after the first year.

PRACTICE PROBLEMS

9.13 To reduce their satellite television expenses, the Johanson family decides to change their programming choice to a plan that is $24.99 per month and eliminate sports from their add-ons. If everything else remains the same, how much will it reduce their monthly expense for satellite television? Compare to the monthly charges after the first year.

9.14 To further reduce their monthly television expenditure, the Johanson family decides to eliminate the movies add-on from their monthly satellite expenditure and instead participate in a movie rental program that costs $14.99 per month. How much should they budget per month for satellite television and movie rental?

In addition to cell phone, Internet, and television services, there are many other areas in which you can reduce expenditures on personal expenses. Many times it takes redefining expectations, finding less expensive ways to do and have the things you want, or going without certain items or services until you can afford them.

Chapter

10

Business Costs

There are many expenses associated with any business. A business that expects to be profitable must be able to plan for and manage the costs associated with doing business.

10.1 Payroll Costs

For most businesses, payroll costs and the direct cost of goods sold are the greatest business expenses. For a service business, payroll is the single greatest business cost.

In Chapters 2 and 3, you calculated wages and deductions from a worker's pay, including federal income tax and FICA tax. An employer is also responsible for paying FICA tax as well as taxes for the Federal Unemployment Tax Act (FUTA) and their State Unemployment Tax Act (SUTA).

Historically, both the employer and employee were each responsible for paying 6.2% of an employee's wages up to a certain wage limit for Social Security tax and 1.45% for Medicare tax, making both responsible for a combined rate of 7.65% for FICA tax. For the year 2011, the Social Security tax rate for employees was reduced to 4.2% on the first $106,800 of wages. This reduction did not affect the employer's portion of the Social Security tax, so the employer's share of FICA tax remained at 7.65%.

Both FUTA and SUTA taxes are only assessed on the first $7,000 of a worker's income per year. FUTA tax rates are 6.2% on the first $7,000 of income. However, an employer can take up to a 5.4% credit on the SUTA taxes that they pay. If the SUTA tax rate is 5.4% or higher, the FUTA rate is reduced to 0.8%. If the SUTA rate is less than 5.4%, then the FUTA rate is reduced by the percent of the SUTA taxes.

Example: Blanchard Industries hires a new manager at an annual salary of $48,000. If the SUTA tax rate is 5.4%, calculate the employer's taxes on the worker's first month's salary.

Solution: Divide the annual salary by 12 to find the monthly salary. Subtract the SUTA rate from the maximum FUTA rate of 6.2%. Multiply the monthly salary by the FICA tax rate of 7.65%, the SUTA rate of 5.4% and the calculated FUTA rate. Add the FICA, FUTA and SUTA taxes to find the total employer's taxes.

Monthly salary = $48,000 ÷ 12 = $4,000

FUTA rate = 6.2% − 5.4% = 0.8%

FICA taxes = $4,000 × 7.65% = $4,000 × 0.0765 = $306

FUTA taxes = $4,000 × 0.8% = $4,000 × 0.008 = $32

SUTA taxes = $4,000 × 5.4% = $4,000 × 0.054 = $216

Employer's taxes = $306 + $32 + $216 = $554

The total employer's taxes paid on the first month's salary are $554.

PRACTICE PROBLEMS

10.1 From the preceding example, what are the employer's taxes for the worker's second month's salary?

10.2 What are the employer's taxes for the worker for the entire year?

In addition to wages and taxes, employers must consider the cost of employee benefits. Health insurance premiums paid by the employer, retirement programs, bonuses, and other benefits are all provided to employees at the employer's expense.

Example: The Hampton Beach Resort employs Bernice for an annual salary of $36,000. The employer must pay 7.65% for FICA taxes, 4.8% for SUTA taxes, and 6.2% FUTA taxes, less the credit for SUTA taxes. In addition, the Hampton Beach Resort provides a health insurance benefit of $150 per month, a retirement benefit equal to 7% of Bernice's salary, and a 2% bonus after completing 11 months of continuous employment. What is the total cost to the employer to employ Bernice for one year?

Solution: Multiply the yearly salary by the FICA rate, the bonus percent, and the retirement contribution percent. Multiply the FUTA and SUTA rates by the $7,000 wage limit. Multiply the monthly health insurance benefit by 12 to find the annual cost. Add the taxes, benefits, and the annual salary to find the total cost for one year.

FICA taxes = $36,000 × 7.65% = $36,000 × 0.0765 = $2,754

Bonus = $36,000 × 2% = $36,000 × 0.02 = $720

Retirement contribution = $36,000 × 7% = $36,000 × 0.07 = $2,520

FUTA tax rate = 6.2% − 4.8% = 1.4%

FUTA tax = $7,000 × 1.4% = $7,000 × 0.014 = $98

SUTA tax = $7,000 × 4.8% = $7,000 × 0.048 = $336

Health insurance = $150 × 12 = $1,800

Total cost = $36,000 + $2,754 + $720 + $2,520 + $98 + $336 + $1,800 = $44,228

The total cost to employ Bernice for one year is $44,228.

Many employees work for an employer for several years or more, and the cost for an employee will typically increase from year to year as employee's wages are increased for outstanding work, loyalty to the company, or cost-of-living adjustments (COLA). Employers may provide a COLA wage increase to all employees in order for employees' wages to keep up with inflation. COLA adjustments will typically be applied as a percentage increase to an employee's wages.

Example: From the previous example, Bernice receives a 3.5% COLA for her second year of employment. If she is also eligible for a 2.5% bonus, and the other benefits and employer taxes apply, what is the total cost to employ Bernice the second year?

Solution: Multiply last year's salary from the previous example by the COLA rate, and add the result to the previous salary to find the new salary. Multiply the yearly salary by the FICA rate, the bonus percent, and the retirement contribution percent. Multiply the FUTA and SUTA rates by the $7,000 wage limit. Multiply the monthly health insurance benefit by 12 to find the annual cost. Add the taxes, benefits, and the annual salary to find the total cost for one year.

COLA = $36,000 × 3.5% = $36,000 × 0.035 = $1,260

New salary = $36,000 + $1,260 = $37,260

FICA taxes = $37,260 × 7.65% = $37,260 × 0.0765 = $2,850.39

Bonus = $37,260 × 2.5% = $37,260 × 0.025 = $931.50

Retirement contribution = $37,260 × 7% = $37,260 × 0.07 = $2,608.20

FUTA tax rate = 6.2% − 4.8% = 1.4%

FUTA tax = $7,000 × 1.4% = $7,000 × 0.014 = $98

SUTA tax = $7,000 × 4.8% − $7,000 × 0.048 = $336

Health insurance = $150 × 12 = $1,800

Total cost = $37,260 + $2,850.39 + $931.50 + $2,608.20 +
$98 + $336 + $1,800 = $45,884.09

The total cost to employ Bernice for the second year is $45,884.09.

PRACTICE PROBLEMS

10.3 C&J Sales wants to give their employees a 4% COLA. If they currently pay $365,000 annually in wages, what will be the annual expense for wages after the COLA?

10.4 How will a COLA affect the amount an employer pays for FICA, FUTA, and SUTA taxes?

10.2 Property and Office Costs

For many businesses, the cost of business space, whether it is an office space, retail store, or a manufacturing facility, is another large expense. Expenses related to business space can include rent, utilities, insurance, cleaning, and maintenance. Purchasing business space will generate other expenses, such as taxes, additional insurance, and mortgage interest.

A business may analyze departments or units of the business by the cost to run that specific department or unit. Expenses, such as rent or utility expenses, may be divided up based on the amount of floor space a department occupies. The payroll costs for a manager's position may be divided up over the number of workers the manager supervises. Taxes and insurance may be divided up based on the value of equipment in each department.

Example: Jameson Manufacturing Company pays $46,500 per month to rent its manufacturing facility. The company charges rent to each department based on the floor space the department occupies in the building. Department 1 has 3,000 square feet. Department 2 has 4,500 square feet, and Department 3 has 8,000 square feet. What amount of the rent should be charged to each department?

Solution: Find the total floor space by adding the square footage that each division occupies. Find the cost per square foot by dividing the rent by the total square footage. To find the amount charged per department, multiply the cost per square foot by the square footage of each department.

Total floor space = 3,000 + 4,500 + 8,000 = 15,500 square feet

Cost per square foot = $46,500 ÷ 15,500 sq. ft. = $3 per square foot

Department 1 rent = 3,000 × $3 = $9,000

Department 2 rent = 4,500 × $3 = $13,500

Department 3 rent = 8,000 × $3 = $24,000

Department 1's share of the rent is $9,000. Department 2's share of the rent is $13,500, and Department 3's share of the rent is $24,000.

PRACTICE PROBLEMS

10.5 A 5,000-square-foot office building rents for $70,000 per year. If utilities and other costs associated with the building are $10,000, how much does a single workstation that occupies 75 square feet cost per year?

10.6 A manufacturing facility costs $250,000 for rent and utilities per year. If the office occupies 10% of the facility, what are the monthly rent and utility costs for the office space?

Unit Cost of Office Work

To analyze the cost of office work, a business may calculate the cost of a single job, a workstation, or some unit of work, such as the cost to produce a report. These unit costs can be compared to industry standards and averages to evaluate efficiency.

Example: Four workstations in an office cost the following amounts last year: salaries and benefits, $100,000; rent and utilities, $3,800; equipment depreciation, $8,500; and supplies, $6,000. What was the average cost per workstation last year?

Solution: Add the cost to find the total yearly cost for the four work stations. Divide the total cost by the number of workstations.

Total cost = $100,000 + $3,800 + $8,500 + $6,000 = $118,300

Cost per workstation = $118,300 ÷ 4 = $29,575

The average cost per workstation last year was $29,575.

PRACTICE PROBLEMS

10.7 To produce 5,000 copies of a report required the following expenses: employee costs, $250; supplies, $5,500; and equipment and other expenses, $480. What was the cost per report?

10.8 A worker in the human resources department of a company processes applications for employment. The worker spends an average of 8 minutes on each application. If it costs the company $15.50 per hour to employ the worker, including wages, taxes, and benefits, what is the cost to process each application?

10.3 Manufacturing Costs

Businesses that manufacture goods use raw materials to make goods to sell. The costs associated with manufacturing include raw materials, direct labor, and factory overhead. *Raw materials* costs are the costs of the materials used to make the products. *Direct labor* costs are the wages and benefits provided to the workers whose job it is to manufacture the products. *Factory overhead* costs are the costs to keep the business running, including wages and benefits for workers who do not directly produce the products, such as office workers or managers, building and equipment expenses, and factory supplies.

The *prime cost* of manufacturing is the sum of the raw materials and the direct labor costs. The *total manufacturing cost* is the sum of the prime cost and the factory overhead.

- Prime Cost = Raw Materials + Direct Labor
- Total Manufacturing Cost = Prime Cost + Factory Overhead

Example: The costs of manufacturing 10,000 units of a certain toy include $9,000 for raw materials, $12,000 for direct labor, and $3,000 for factory overhead. What are the prime cost and total manufacturing cost for producing the toys?

Solution: Add the costs of the raw materials and direct labor to find the prime costs. Add the prime cost to the factory overhead to find the total manufacturing cost.

Prime cost = $9,000 + $12,000 = $21,000

Total manufacturing cost = $21,000 + $3,000 = $24,000

The prime cost is $21,000 and the total manufacturing cost is $24,000 to manufacture the 10,000 toys.

Practice Problems

10.9 The costs to make 25,000 necklaces are raw materials, $85,000; direct labor, $10,800; and factory overhead, $2,500. What is the prime cost of the necklaces? What is the total manufacturing cost?

10.10 In the previous problem, what is the manufacturing cost per necklace?

Breakeven Point

Prime costs are generally classified as *variable costs* because they vary with the amount of products produced. To produce more, you need more raw materials and more total hours from direct labor. Factory overhead costs are generally classified as *fixed costs* because they do not change with the amount of products produced.

The *breakeven point* is the point at which the cost to produce and sell goods is equal to the income from sales. The breakeven point is affected by both variable and fixed costs.

To calculate the breakeven point, divide the fixed costs by the difference of the sales price per unit and the variable cost by unit.

$$\text{Breakeven Point} = \frac{\text{Fixed Cost}}{\text{Sales Price per Unit} - \text{Variable Cost per Unit}}$$

Example: C & J Manufacturing Corporation produces display boxes for flags. They sell the boxes for $20 each. The estimate for fixed costs is $15,000. The variable costs for producing each display box are $8. How many display boxes must be sold to break even?

Solution: Substitute the fixed costs, sales price, and variable cost per unit in the breakeven point formula and evaluate.

$$\text{Breakeven point} = \frac{\$15,000}{\$20 - \$8} = \frac{\$15,000}{\$12} = 1,250$$

C & J must sell 1,250 display boxes to break even.

PRACTICE PROBLEMS

10.11 Silk Screening Incorporated plans to sell a new line of silk screened t-shirts for $25 each. They estimate fixed expenses of $45,000 and variable expenses of $13 per shirt. How many t-shirts must be sold to break even?

10.12 Silk Screening Incorporated found a different supplier for the t-shirts they will silk screen. The variable expenses to produce the t-shirts will decrease to $10 per shirt. How many t-shirts must be sold to break even?

10.4 Depreciation Costs

A major expense for businesses is the depreciation of property and equipment. Depreciation can be calculated in a number of ways, including the declining balance method, the sum-of-the-years-digits method, and the modified accelerated cost recovery system.

The *book value* is the original cost of the property minus the depreciation.

Declining Balance Method

The *declining balance* method uses a fixed rate of depreciation per year. Since the constant rate is applied to a declining book value, the amount of depreciation decreases over the life of the property.

Example: A truck used by a business for deliveries was purchased for $18,000. It is estimated to depreciate at the rate of 20% per year. What are the book values at the end of the first three years?

Solution: Multiply the original cost of the truck by the rate of depreciation. Subtract the depreciation from the original cost of the truck to find the book value at the end of the first year. Multiply the book value by the rate of depreciation. Subtract the depreciation from the book value to find the new book value. Repeat for the third year.

First year depreciation = $18,000 × 20% = $18,000 × 0.20 = $3,600

Book value at end of first year = $18,000 − $3,600 = $14,400

Second year depreciation = $14,400 × 20% = $14,400 × 0.20 = $2,880

Book value at the end of second year = $14,400 − $2,880 = $11,520

Third year depreciation = $11,520 × 20% = $11,520 × 0.20 = $2,304

Book value at the end of third year = $11,520 − $2,304 = $9,216

The book values at the end of the first three years are $14,400, $11,520, and $9,216.

PRACTICE PROBLEM

10.13 Stonewall Industries purchased a new piece of equipment for $25,000. The estimated rate of depreciation is 15%. What are the book values at the end of the first three years using the declining balance method?

Sum-of-the-Years-Digits Method

Another depreciation method that takes the greatest amount of depreciation in the early years is the *sum-of-the-years-digits method*. The sum-of-the-years-digits method uses a variable rate to calculate the depreciation.

To use this method, estimate the number of years you will use the property or equipment. Estimate the value at the end of the number of years and subtract from the original cost to find the total expected depreciation.

Next, add the sum of the digits from 1 to the number of years of use estimated. For example, if you estimate that a piece of equipment will be used for four years, you add, $1 + 2 + 3 + 4 = 10$ years. The first year, you depreciate the equipment $\frac{4}{10}$ of the total expected depreciation. In the second year, you depreciate $\frac{3}{10}$ of the total expected depreciation, $\frac{2}{10}$ for the third year, and $\frac{1}{10}$ for the fourth year.

Example: A truck used by a business for deliveries was purchased for $18,000. The business expects to use the truck for five years, and then trade it in for $6,000. What are the book values at the end of the first three years?

Solution: Find the total expected depreciation by subtracting the trade-in value from the purchase price. Find the sum-of-the-years-digits by adding the digits 1 through 5. Find the depreciation for each year by multiplying the total expected depreciation by the depreciation fraction. Subtract the depreciation from the purchase price the first year, and the previous year's book value after the first year, to find the new book value.

Total expected depreciation = $18,000 − $6,000 = $12,000

Sum-of-the-years-digits = 1 + 2 + 3 + 4 + 5 = 15 years

Depreciation for the first year = $12,000 × $\dfrac{5}{15}$ = $4,000

Book value at the end of the first year = $18,000 − $4,000 = $14,000

Depreciation for the second year = $12,000 × $\dfrac{4}{15}$ = $3,200

Book value at the end of the second year = $14,000 − $3,200 = $10,800

Depreciation for the third year = $12,000 × $\dfrac{3}{15}$ = $2,400

Book value at the end of the third year = $10,800 − $2,400 = $8,400

The book values at the end of the first three years are $14,000, $10,800 and $8,400.

PRACTICE PROBLEM

10.14 Stonewall Industries purchased a new piece of equipment for $25,000. After seven years, they plan to sell the equipment for $8,000. What are the book values at the end of the first three years using the sum-of-the-years-digits method?

Modified Accelerated Cost Recovery System Method (MACRS)

For federal income tax purposes, businesses must use the modified accelerated cost recovery system (MACRS) for most business property put into service after 1986. The MACRS system has a specific percent of depreciation that can be taken based on which class the property falls into.

Table 10.1 shows the depreciation allowed for two classes of property based on a five-year and seven-year span of use. Notice that the five-year and seven-year classes depreciate the property over six and eight years, respectively, because there is an assumption that the property was acquired in the middle of the tax year, qualifying for half a year's worth of depreciation in the first year and in the last year.

Table 10.1 MACRS Depreciation Table

Class Life	Cars, Trucks, and Office Equipment 5 years	Office Furniture and Fixtures 7 years
First Year	20.0%	14.29%
Second Year	32.0%	24.49%
Third Year	19.2%	17.49%
Fourth Year	11.52%	12.49%
Fifth Year	11.52%	8.93%
Sixth Year	5.76%	8.92%
Seventh Year		8.93%
Eighth Year		4.46%

To find the depreciation for a given year, find the appropriate percent using the year of service and the class. Multiply the depreciation percent by the original cost of the property.

Example: A truck used by a business for deliveries was purchased for $18,000. What are the book values at the end of the first three years?

Solution: Identify which class the truck will fall into. For each year of depreciation, multiply the percent from the table by the original value of the truck. To find the book value, subtract the depreciation from the original value the first year, and subtract the depreciation from the previous book value for every year after the first.

   The truck is in the five-year class.

   Depreciation for the first year = $18,000 × 20.0% = $18,000 × 0.20 = $3,600

   Book value at the end of the first year = $18,000 − $3,600 = $14,400

   Depreciation for the second year = $18,000 × 32.0% = $18,000 × 0.32 = $5,760

   Book value at the end of the second year = $14,400 − $5,760 = $8,640

   Depreciation for the third year = $18,000 × 19.2% = $18,000 × 0.192 = $3,456

   Book value at the end of the third year = $8,640 − $3,456 = $5,184

   The book values at the end of the first three years are $14,400, $8,640, and $5,184.

Practice Problem

10.15 Stonewall Industries purchased a new piece of equipment for $25,000. The equipment has a class life of seven years. What are the book values at the end of the first three years using MACRS?

10.5 Shipping Costs

Shipping costs can affect businesses both in acquiring raw materials and other supplies and in getting products to their customers. Businesses may use the U.S. postal service, private carriers, and other shippers. Carriers will have different restrictions on the size and weight of packages, and many will offer additional services, such as faster delivery service, delivery confirmation, or insurance.

For many carriers, the two most important factors in calculating the cost to ship a package are the weight and the destination zone. Table 10.2 shows a sample price chart for a carrier.

Table 10.2 Sample Charges for 3-4 Day Shipping Service

Weight Up To (lb)	Destination Zone						
	1 & 2	3	4	5	6	7	8
1	$4.80	$4.85	$4.95	$5.05	$5.15	$5.24	$5.44
2	$4.90	$4.99	$5.38	$6.82	$7.41	$7.90	$8.82
3	$5.08	$5.86	$6.79	$8.19	$9.35	$10.07	$11.74
4	$5.72	$6.76	$7.78	$10.14	$11.89	$12.68	$14.14
5	$6.67	$7.84	$8.87	$11.76	$13.57	$14.57	$16.37
6	$7.35	$6.64	$9.96	$13.40	$15.22	$16.62	$18.71
7	$7.94	$9.58	$11.14	$15.23	$16.82	$18.76	$21.04

Example: Coastal Aviation is shipping products to a customer. The box weighs 5 pounds and has a declared value of $200. The shipping carrier uses the rates from Table 10.2. The carrier charges $1.40 per $100 of declared value for insurance. How much will it cost to insure and ship the box to a customer in zone 6?

Solution: Identify the shipping charges from Table 10.2 by finding the charge in the row representing 5 pounds and the column for destination zone 6. To find the cost for insurance, divide the declared value by $100 and multiply the result by $1.40. Add the shipping and insurance cost together to find the total cost.

Shipping charge from Table 10.2 = $13.57

$$\text{Insurance} = \frac{\$200}{\$100} \times \$1.40 = 2 \times \$1.40 = \$2.80$$

Total cost = $13.57 + $2.80 = $16.37

The cost to insure and ship the product to the customer is $16.37.

PRACTICE PROBLEMS

10.16 Using the shipping rates in Table 10.2, how much will it cost to ship three boxes that weigh 2 pounds each to zone 4?

10.17 What is the total cost to ship a 4-pound box to zone 7 if insurance costs $1.40 per $100 of declared value and the declared value is $500? Use Table 10.2.

Larger, bulkier items will typically be shipped via *freight*. Freight shipments may be transported via airplane, truck, train, barge, or ship.

The term *f.o.b.* stands for free on board or freight on board, and it is used to designate which party, the buyer or the seller, is paying for the shipping costs and is legally responsible for the goods in transit. The transportation and liability for goods that are designated as f.o.b. shipping point lie with the buyer; the buyer is responsible for the shipping costs and liability for the goods from the shipping point. The transportation and liability for goods that are designated f.o.b. destination lie with the seller; the seller is responsible for the shipping costs and liability for the goods to the destination point.

Shipping companies may use a rate chart similar to Table 10.2 to calculate freight charges, or they may charge a certain fee per 100 pounds, or *cwt* (*hundredweight*). To find the charge using a rate per cwt, divide the weight of the freight by 100 and then multiply by the rate per cwt.

Example: Creekstone Tool and Die Company is shipping a 10,000-pound piece of equipment from their factory in Dallas to a customer in Toledo. The terms are f.o.b. Toledo. The freight charges are $45.87 cwt. How much are the shipping charges, and who is responsible for the equipment in transport?

Solution: Divide the weight by 100 and multiply by $45.87 to find the freight charges.

The terms are f.o.b. destination, so the seller is responsible for the freight.

$$\text{Freight charges} = \frac{10,000}{100} \times \$45.87 = 100 \times \$45.87 = \$4,587$$

The freight charges are $4,587, and the seller is responsible for the charges and shipment to the buyer.

PRACTICE PROBLEMS

10.18 The Crocket Company is shipping 4,000 pounds of freight from Chicago to Milwaukee, f.o.b. Chicago. The freight charge is $10.75 a cwt. What is the freight charge, and who is responsible for paying the charge?

10.19 Company A in San Francisco offers to ship 2,000 pounds of raw materials to your company in Sacramento for $8.25 a cwt, f.o.b. San Francisco. Company B offers to ship 2,000 pounds of raw materials to your company in Sacramento for $12.25 per cwt, f.o.b. Sacramento. Which company is offering the best shipping terms for your company?

10.6 Travel Expenses

When employees travel for business, whether it is an isolated trip or a regular part of an employee's job, an employer will often reimburse an employee's *travel expenses*. Travel expenses may include airline tickets, hotel accommodations, meals, rental car charges, entertainment expenses for entertaining clients, and mileage driven in a personal vehicle.

An employer may reimburse employees when they use their personal vehicle for business purposes. Many employers reimburse employees the same rate per mile that is allowed by the IRS for tax purposes.

Example: Joanna uses her car while making sales calls for her employer. Her employer reimburses her $0.50 per mile. Last year, Joanna drove her car 875 miles on company business. How much was Joanna reimbursed for mileage?

Solution: Multiply the number of business miles driven by the reimbursement rate.

Reimbursement = 875 × $0.50 = $437.50

Joanna should be reimbursed $437.50.

PRACTICE PROBLEMS

10.20 If Bernard drives an average of 700 miles per month in his personal car, and his employer reimburses him at a rate of $0.52 per mile, how much should Bernard expect to receive in mileage reimbursement for the year?

10.21 Francesca drove her car 25,000 miles last year. Sixty percent of the miles she drove last year were for business purposes, and her employer reimburses at the rate of $0.50 per mile. How much will Francesca receive in mileage reimbursement?

When an employer sends an employee on a business trip, the employer will usually pay for the cost of travel and for housing and meals while the employee is on the trip. Typically an employer will pay employees for their actual expenses, while others may pay a per diem, or per day rate, regardless of what the employee chooses to spend.

Example: Paul's employer requires him to attend a continuing education conference each year. The employer paid the $250 registration fee to the organization hosting the conference. Paul purchased an airline ticket for $382 that will be reimbursed by his employers. In addition, the employer will pay $150 per diem for the five-day conference. How much will Paul's employer reimburse him for the trip?

Solution: Multiply the per diem rate by five days. Add the per diem to the cost of the airfare to find the total reimbursement.

Per diem = $150 × 5 = $750

Reimbursement = $750 + $382 = $1,132

Paul's employer will reimburse him $1,132 for the conference.

PRACTICE PROBLEMS

10.22 Nate attended a continuing education seminar for his job. His employer reimbursed him $345 for airfare, $225 for hotel expenses, $65 for a rental car, and $75 for meals. What was Nate's total reimbursement?

10.23 Shelly attended a four-day conference last week. She drove 500 miles round trip to the conference and paid the $150 conference fee. She spent $129 per night for three nights in a hotel. Her employer will reimburse her $0.55 per mile for the miles driven to the conference, the fee for the conference, hotel charges, and $30 per day for meals. How much will Shelly receive in reimbursement?

Chapter

11

Sales

B usinesses make financial transactions with suppliers and customers by cash or credit. Tracking those transactions, as well as pricing merchandise for sale, can be important parts of running a successful business.

11.1 Sales Slips

A *sales slip* records the sale of goods or services. Many businesses use computer- or cash-register-generated sales slips, although other businesses use hand-recorded sales slips. A sales slip will record all of the items purchased or services provided, the price of each, a subtotal, sales tax (if applicable), and the total amount of the purchase.

Example: Sharon works at a boutique. She sold the following items to a customer: two shirts for $35 each, one pair of shorts for $45, and one pair of shoes for $23. If the sales tax rate is 5%, what is the total amount of the sale?

Solution: Multiply the cost of a shirt by 2, and add the other items to find the subtotal. Multiply the subtotal by 5% to find the amount of sales tax. Add the subtotal and the sales tax to find the total amount of the purchase.

Cost of 2 shirts = $35 × 2 = $70

Subtotal = $70 + $45 + $23 = $138

Sales tax = $138 × 5% = $138 × 0.05 = $6.90

Total = $138 + $6.90 = $144.90

The total amount of the sale is $144.90.

If an item is on sale, the sale price of the item needs to be calculated before finding the subtotal or sales tax due.

Example: Jack works in a hardware store. A customer purchases a broom and dustpan for $15.99 and a tool set that is on sale for 35% off. If the original price of the tool set was $59, and the sales tax rate is 4%, what is the total amount due for the sale?

Solution: Multiply the original price of the tool set by the discount to find the discount amount. Subtract the discount amount from the original price to find the sale price. Add the cost of the tool set and broom and dustpan to find the subtotal. Multiply the subtotal by the sales tax rate. Add the subtotal and the sales tax to find the total amount due.

Discount = $59 × 35% = $59 × 0.35 = $20.65

Sale price = $59 − $20.65 = $38.35

Subtotal = $15.99 + $38.35 = $54.34

Sales tax = $54.34 × 4% = $54.34 × 0.04 = $2.17

Total = $54.34 + $2.17 = $56.51

The total amount due is $56.51.

PRACTICE PROBLEMS

11.1 Jason sold the following items to a customer: three pairs of bike gloves for $18.25 each, two bike helmets for $23.99 each, and one water bottle for $5.62. If the sales tax rate is 5%, what is the total amount of the sale?

11.2 Sammy purchases a trashcan for $9.99 and an area rug that is on sale for 25% off. The original price of the area rug was $63.49. If the sales tax rate is 3.5%, how much does Sammy owe?

11.2 Managing Cash

During the course of a business day, many businesses conduct cash transactions. Typically at the end of each business day, the money in the cash register will be counted to compare against the transaction records and verify that the correct amount of money is in the cash register. The process is called *proving cash*. If you have less cash than you should according to the transaction records, you are *cash short*. If you have more cash than you should, you are *cash over*.

Example: Becca works for a Heaven Scent Gift Shop. She began the day with $200 in her cash register. During the course of the day, she received $3,255.81 for cash sales. She paid out $33.25 in cash for a refund. When she counted the cash in the register at the end of the day, she had $3,423.08. Was she cash over or cash short? How much was she short or over?

Solution: Add the beginning cash amount to the cash receipts. Subtract the cash paid out to find the amount of cash that should be in the register. Compare that amount to the amount she counted in the drawer to determine if she is cash short or cash over. Subtract the smaller amount from the larger amount to find out by how much she is short or over.

Total cash in = $200 + $3,255.81 = $3,455.81

Amount that should be in register = $3,455.81 − $33.25 = $3,422.56

The amount in the register is greater than what should be there, so Becca is cash over.

$3,423.08 − $3,422.56 = $0.52

The register is cash over by $0.52.

PRACTICE PROBLEMS

11.3   Nicholas works in a garden shop. At the beginning of his shift, the cash register had $250 in it. His sales totaled $1,285.36. He paid out $146.92. He had $1,387.29 in the register at the end of his shift. Was Nicholas cash short or cash over? How much was he short or over?

11.4   Janice started the workday with $100 in the cash register. During the business day, she received $2,486.10. She paid out $239.86. At the end of the day, the register had $2,346.24 in the register. Is the cash register cash over or cash short? By how much?

11.3 Invoices and Credit

Goods may also be purchased on *credit* when the buyer and seller have an arrangement for the buyer to pay the seller at a later date. A wholesaler may sell goods to a retailer on credit, and a retailer may sell goods to its customers on credit. When goods are sold on credit, a *sales invoice* documents the goods sold and the total amount due.

If goods purchased on credit are returned, the customer's account is credited the amount of the return, and a *credit memo* documents the transaction. The business that has extended credit must keep an accurate accounting of their customers' account balances.

Example: The Office Supply Place purchases goods for their retail store from Office Wholesale Club on credit. Over the last month, The Office Supply Place made two purchases from Office Wholesale Club: Invoice 1025 for $1,568.22 and Invoice 1038 for $5,729.38. The Office Supply Place returned $210.45 worth of defective merchandise recorded on Credit Memo 58, and made a payment of $8,000 on their account. If the beginning balance at the first of the month was $4,822.19, what is the balance at the end of the month?

Solution: Add the invoices to find the total amount of charges to the account. Add the payment and credit invoice to find the total amount of credits to the account. To the beginning balance, add the charges and subtract the credits.

Charges = $1,568.22 + $5,729.38 = $7,297.60

Credits = $210.45 + $8,000 = $8,210.45

Balance = $4,822.19 + $7,297.60 − $8,210.45 = $3,909.34

The Office Supply Place's new balance is $3,909.34.

PRACTICE PROBLEMS

11.5 At the beginning of the month, the account for The Fashion Place had a balance of $1,248.62. The sales to The Fashion Place were recorded on Invoice #2293 for $842.98 and Invoice #2304 for $622.18. The Fashion Place made a payment on its account in the amount of $2,000. What was the account balance for The Fashion Place at the end of the month?

11.6 Marine Motor Products allows customers to buy on store credit. Joe has an account at Marine Motor Products with a balance of $322.18. Joe purchases supplies at the 20% storewide sale. The items have a total selling price of $155 before the 20% discount. If there is a 5.5% sales tax, and Joe charges the items to his account, what is the total of his account after the purchase?

Cash Discounts

Many businesses will offer their retail customers discounts for paying invoices early. The discount is figured on the *invoice price*, the total due on the invoice. The terms for a cash discount may be stated as "2/10, 1/30, n/60." This means that 2% can be deducted from the total due if the invoice is paid within 10 days of the invoice date, 1% can be deducted from the total due if the invoice is paid within 30 days of the invoice date, and the balance is due within 60 days of the invoice date.

Another identifier that may be used is EOM, meaning end of the month. So payment terms of 2/10 EOM means that 2% can be deducted from the invoice price if the balance is paid within 10 days after the end of the month shown on the invoice.

To find the *cash price* after a discount, first multiply the invoice price by the rate of the cash discount to find the amount of discount, and then subtract the cash discount from the invoice price.

- Cash Discount = Invoice Price × Rate of Cash Discount
- Cash Price = Invoice Price − Cash Discount

Shipping costs are not subject to the cash discount, so any applicable shipping costs must be added to the cash price to find the total due. Remember that f.o.b destination means the seller will pay the shipping charges, while f.o.b. shipping point or origin means the buyer will pay the shipping charges.

Example: The Green Grocer in Tacoma, WA, purchased produce from Organic Growers in Spokane, WA. The invoice, dated July 10, states terms of 2/10, 1/30, n/60. The invoice price is $2,340. The shipping charges are $125 f.o.b. Spokane. If the Green Grocer pays the invoice by July 22, what is the amount due?

Solution: Determine the dates for the cash discounts by adding 10 days and 30 days to the invoice date. Identify the percent of cash discount. Multiply the invoice price by the cash discount percent to find the cash discount. Subtract the cash discount from the invoice price to find the cash price. Determine who pays the shipping charges and add to the cash price if the buyer is responsible.

2% discount period = July 10 + 10 days = July 20

1% discount period = July 10 + 30 days = July 31 + 9 days = August 9

The 1% cash discount applies.

Cash discount = $2,340 × 1% = $2,340 × 0.01 = $23.40

Cash price = $2,340 − $23.40 = $2,316.60

Shipping terms are f.o.b. shipping point, so the Green Grocer is responsible for the shipping charges.

Amount due = $2,316.60 + $125 = $2,441.60

On July 22, the Green Grocer should pay $2,441.60 to Organic Growers.

PRACTICE PROBLEMS

11.7 Simpson's Hardware purchases paint from a wholesaler. The amount due on a recent invoice is $5,480.62. The wholesaler offers a cash discount of 2/10 EOM. If the invoice is paid during the cash discount period, how much is due?

11.8 The terms on an invoice dated February 1 are 2/10, 1/30, n/60. The invoice total is $9,462.89 with shipping charges of $225 f.o.b. destination. If the invoice is paid on February 9, what is the total amount due?

Rate of Cash Discount

If you know both the invoice price and the cash price, you can calculate the rate of discount given by dividing the cash discount by the invoice price and changing the result to a percent.

$$\text{Rate of Cash Discount} = \frac{\text{Cash Discount}}{\text{Invoice Price}}$$

Example: The Gift Shoppe received a cash discount and paid $1,176 to pay off an invoice price of $1,200. What is the rate of the cash discount applied to the invoice price?

Solution: Subtract the cash price from the invoice price to find the amount of the cash discount. Divide the cash discount by the invoice price and change to a percent by multiplying by 100 or moving the decimal point two places to the right.

Cash discount = $1,200 − $1,176 = $24

$$\text{Rate of cash discount} = \frac{\$24}{\$1,200} = 0.02 = 2\%$$

The Gift Shoppe was given a 2% cash discount.

Practice Problems

11.9   A business takes advantage of a cash discount offered. If the business paid $997.50 to settle an invoice for $1,050, what was the rate of cash discount given?

11.10   An invoice for $7,500 was paid during a cash discount period. If the amount paid was $7,237.50, what were the amount of cash discount and the rate of cash discount?

11.4 Trade Discounts

Trade discounts are list price reductions that a business may offer to its business customers. A trade discount is calculated on the list price per item, rather than on the total amount due, because different trade discounts may be applied to different items.

To calculate the trade discount, multiply the list price by the rate of trade discount. The invoice price for that item is the list price minus the trade discount.

- Trade Discount = List Price × Rate of Trade Discount
- Invoice Price − List Price − Trade Discount

Example: The list price of a dress is $85. The trade discount given to a clothing retailer is 40%. What invoice price will a retailer pay for 50 dresses?

Solution: Multiply the list price by 40% to find the discount. Subtract the discount from the list price to find the invoice price. Multiply the invoice price by the number of dresses purchased.

Trade discount = $85 × 40% = $85 × 0.40 = $34

Invoice price for 1 dress = $85 − $34 = $51

Invoice price for 50 dresses = $51 × 50 = $2,550

The invoice price of 50 dresses is $2,550.

PRACTICE PROBLEMS

11.11 The list price of a knife set is $75 with a 30% trade discount to retailers. What trade discount is given to retailers?

11.12 A desk has a list price of $399. What is the invoice price for a retailer who gets a $35\frac{1}{2}$% trade discount?

Rate of Trade Discount

The rate of trade discount can be calculated by dividing the discount by the list price.

$$\text{Rate of Trade Discount} = \frac{\text{Trade Discount}}{\text{List Price}}$$

Example: A bicycle has a list price of $248. The bicycle is sold to a retailer for an invoice price of $161.20. What rate of trade discount was given?

Solution: Subtract the invoice price from the list price to find the trade discount. Divide the trade discount by the list price and change to a percent by multiplying by 100 or moving the decimal point two places to the right.

Trade discount = $248 − $161.20 = $86.80

$$\text{Rate of trade discount} = \frac{\$86.80}{\$248} = 0.35 = 35\%$$

A 35% trade discount was given to the retailer.

PRACTICE PROBLEMS

11.13   A lawn chair with a list price of $20 was sold to a retailer for $11. What rate of trade discount was given?

11.14   A retailer purchases 150 scarves for $9,300. If the list price of each scarf is $95.38, what trade discount was given on the scarves?

Series Discounts

Some businesses will state a trade discount as a *discount series*, or a series of discounts instead of a single trade discount. For example, the trade discount might be 40%, 10%, 5%, meaning all three discounts are given on the item. The discounts might also be written as 25/15/5, meaning a 25%, 15%, 5% series discount.

The first discount is given on the list price. The second discount is given on the remainder that is left after the first discount. The third discount is given on the remainder after the second discount.

Example: The list price of a copper pipe is $150. If the trade discount is 40%, 10%, 5%, what is the invoice price?

Solution: Multiply the list price by the first discount. Subtract the discount from the list price. Use the result as the remainder and apply the next discount. Repeat for each discount.

First discount = $150 × 40% = $150 × 0.40 = $60

First remainder = $150 − $60 = $90

Second discount = $90 × 10% = $90 × 0.10 = $9

Second remainder = $90 − $9 = $81

Third discount = $81 × 5% = $81 × 0.05 = $4.05

Invoice price = $81 − $4.05 = $76.95

The invoice price of the copper pipe is $76.95.

You can simplify the discount series calculation by calculating the single discount that is equivalent to the series discount. Finding the equivalent single discount will help you calculate the discount more quickly and also allow you to compare discounts from different vendors.

Instead of working with a dollar amount, the calculations use the percents. The list price is represented by 100%. The final remainder represents the single discount. The single discount is applied to the list price to find the discount because the single discount equivalent is the percent that has been taken away by the series of discounts.

Example: What is the single discount equivalent to a series discount of 40%, 10%, 5%? Use the single discount equivalent to calculate the invoice price of a copper pipe with a list price of $150.

Solution: Subtract the discount percent from 100% to find the percent of the list price that remains after the first discount. Multiply the second discount by this percent and subtract the result from the first remainder. Repeat for the third discount. Subtract the final remainder from 100% to find the equivalent single discount.

First remainder = 100% − 40% = 60%

Second discount = 60% × 10% = 0.60 × 0.10 = 0.06 = 6%

Second remainder = 60% − 6% = 54%

Third discount = 54% × 5% = 0.54 × 0.05 = 0.027 = 2.7%

Third remainder = 54% − 2.7% = 51.3%

Single discount equivalent = 100% − 51.3% = 48.7%

Discount on copper pipe = $150 × 48.7% = $150 × 0.487 = $73.05

Invoice price = $150 − $73.05 = $76.95

The single discount equivalent is 48.7%, and the invoice price of the copper pipe is $76.95.

Another method for calculating a series of discounts quickly utilizes the complement of each discount. The discount is what is taken away, and the complement of the discount is the percent that remains. To find the complement of a discount, subtract the discount from 100%. If you subtract each discount from 100% and multiply the results, you will find the percent that the invoice price is of the list price. Simply multiply that percent by the list price to find the invoice price.

Example: Use the complement method to find the invoice price of a copper pipe with a list price of $150 and a series discount of 40%, 10%, 5%.

Solution: Find the complement to each discount by subtracting each discount from 100%. Multiply the complements to find what percent the invoice price is of the list price. Multiply the percent by the list price to find the invoice price.

Complements = 100% − 40% = 60%; 100% − 10% = 90%; 100% − 5% = 95%

Combined percent = 60% × 90% × 95% = 0.60 × 0.90 × 0.95 = 0.513 = 51.3%

Invoice price = $150 × 51.3% = $150 × 0.513 = $76.95

PRACTICE PROBLEMS

11.15 A couch with a list price of $800 has a trade discount of 35%, 15%, 10%. What is the invoice price of the couch?

11.16 What is the single discount equivalent of a 45%, 15%, 5% series discount? What is the discount on a $500 item?

11.5 Markup

Markup is the amount of money that businesses add to the cost of an item so that they can cover their expenses and make a profit. The cost for a business to purchase the item plus the markup is the *selling price*.

Markup Based on Cost

Some businesses use the cost of the item as the base from which to figure the markup. For example, a business might mark up the cost of an item by 40%.

To figure the markup and selling price when the markup is based on the cost, first multiply the cost by the markup percent. The result is the dollar amount of markup. Add the markup to the cost to find the selling price.

- Markup = Cost × Rate of Markup
- Selling Price = Cost + Markup

Example: Find the markup and selling price for a dish set that cost the retailer $28 if the retailer has a 45% markup.

Solution: Multiply the cost by 45% to find the amount of markup. Add the markup to the cost to find the selling price.

Markup = $28 × 45% = $28 × 0.45 = $12.60

Selling price = $28 + $12.60 = $40.60

The selling price for the dish set is $40.60.

Using algebra skills, you can rearrange the formula to find the rate of markup based on cost if you know the cost and the selling price.

$$\text{Markup} = \text{Cost} \times \text{Rate of Markup}$$

Divide both sides of the formula by the cost, and the formula becomes

$$\text{Rate of Markup} = \frac{\text{Markup}}{\text{Cost}}$$

Example: The selling price of a drill that cost $34 is $52.70. What is the markup based on the cost of the item?

Solution: Subtract the cost from the selling price to find the markup. Divide the markup by the cost.

Markup = $52.70 − $34 = $18.70

$$\text{Rate of markup} = \frac{\$18.70}{\$34} = 0.55 = 55\%$$

The drill has a 55% markup based on the cost.

You can also rearrange the formula to find the cost if you know the rate of markup on the cost and the selling price.

$$\text{Markup} = \text{Cost} \times \text{Rate of Markup}$$

Divide both sides of the formula by the rate of markup, and the formula becomes

$$\text{Cost} = \frac{\text{Markup}}{\text{Rate of Markup}}$$

Example: If the markup on an item is $20 and the markup rate is 40% based on the cost, find the cost and the selling price of the item.

Solution: Divide the markup by the rate of markup to find the cost. Add the markup to the cost to find the selling price.

$$\text{Cost} = \frac{\$20}{40\%} = \frac{\$20}{0.40} = \$50$$

Selling price = $50 + $20 = $70

The cost of the item is $50, and the selling price is $70.

PRACTICE PROBLEMS

11.17 Find the markup and selling price for a sewing machine that cost the retailer $225 if the retailer has a 55% markup

11.18 The selling price of a sled that cost $18 is $25.56. What is the markup based on the cost of the item?

11.19 If the markup on an item is $8 and the markup rate is 45% based on the cost, find the cost and the selling price of the item.

Markup Based on Selling Price

Sometimes a business has a selling price in mind. For example, a business may want to carry a line of purses that they can sell for $29.99. In this case, a business may think of the markup as a percentage of the selling price. In other words, to make a profit, the business may need to have a 40% markup on the selling price of the purses. The business will need to calculate the maximum cost they can pay for the purses in order to sell them for $29.99.

To find the cost, multiply the markup by the selling price to find the amount of markup. Subtract the markup from the selling price in order to find the cost.

- Markup = Selling Price × Rate of Markup
- Cost = Selling Price − Markup

Example: The Purse Emporium wants to carry a line of purses that each cost $29.99. They need a 40% markup based on the selling price in order to make a profit. What is the maximum they can pay for the purses in order to sell them at a profit?

Solution: Multiply the selling price by the rate of markup. Subtract the markup from the selling price to find the cost.

Markup = $29.99 × 40% = $29.99 × 0.40 = $12.00

Cost = $29.99 − $12.00 = $17.99

If the Purse Emporium purchases the purses for $17.99, they can sell them for $29.99 and make a profit.

To find the rate of markup based on the selling price, divide both sides of the markup formula by the selling price:

$$\text{Markup} = \text{Selling Price} \times \text{Rate of Markup}$$

becomes

$$\text{Rate of Markup} = \frac{\text{Markup}}{\text{Selling Price}}$$

Example: The selling price of a drill that cost $34 is $52.70. What is the markup based on the selling price of the item?

Solution: Subtract the cost from the selling price to find the markup. Divide the markup by the selling price to find the rate of markup based on the selling price.

Markup = $52.70 − $34 = $18.70

$$\text{Rate of Markup} = \frac{\$18.70}{\$52.70} = 0.355 = 35.5\%$$

The rate of markup based on the selling price is 35.5%.

To find the selling price using the markup and rate of markup based on the selling price, divide both sides of the markup formula by the rate of markup:

Markup = Selling Price × Rate of Markup

becomes

$$\text{Selling Price} = \frac{\text{Markup}}{\text{Rate of Markup}}$$

Example: If the markup on an item is $20 and the markup rate is 40% based on the selling price, find the cost and the selling price of the item.

Solution: Divide the markup by the rate of markup to find the selling price. Subtract the markup from the selling price to find the cost.

$$\text{Selling Price} = \frac{\$20}{40\%} = \frac{\$20}{0.40} = \$50$$

Cost = $50 − $20 = $30

The cost of the item is $30, and the selling price is $50.

PRACTICE PROBLEMS

11.20   The Auto Shop wants to carry a line of batteries that each cost $99.99. They need a 40% markup based on the selling price in order to make a profit. What is the maximum they can pay for the batteries in order to sell them at a profit?

11.21   The selling price of a pair of boots that cost $48 is $96. What is the markup based on the selling price of the item?

11.22   If the markup on an item is $50 and the markup rate is 40% based on the selling price, find the cost and the selling price of the item.

11.6 Markdown

A business may put items on sale in order to get rid of an inventory of items or to entice customers to the store with the hope that they will make other purchases. The amount that the business reduces the price of an item is called the *markdown*.

To find the amount of markdown, multiply the marked price by the rate of markdown. To find the new selling price, subtract the markdown from the marked price.

• Markdown = Marked Price × Rate of Markdown

• New Selling Price = Marked Price − Markdown

Example: A sweater that has a selling price of $80 is on sale for 35% off. What is the sale price of the sweater?

Solution: Multiply the selling price by the rate of the markdown to find the amount of the markdown. Subtract the markdown from the selling price to find the sale price.

   Markdown = $80 × 35% = $80 × 0.35 = $28

   Sale price = $80 − $28 = $52

   The sale price of the sweater is $52.

Example: A wholesaler sells a game for a list price of $20. The wholesaler offers the retailer a trade discount of 40%. The retailer sells the game for a 40% markup on the cost of the game. During the holidays, the retailer puts the game on sale for a 20% discount. What is the sale price of the game? Does the sale price cover the cost the retailer paid for the game from the wholesaler?

Solution: Multiply the list price of the game by the rate of the trade discount to find the trade discount. Subtract the discount from the list price to find the cost of the game for the retailer. Multiply the cost by the rate of markup and add the markup to the cost to find the selling price. Multiply the selling price by the rate of discount on the sale. Subtract the sale discount from the selling price to find the sale price. Compare the sale price and the cost from the wholesaler.

Trade discount = $20 × 40% = $20 × 0.40 = $8

Cost = $20 − $8 = $12

Markup − $12 × 40% = $4.80

Selling price = $12 + $4.80 = $16.80

Discount = $16.80 × 20% = $16.80 × 0.20 = $3.36

Sale price = $16.80 − $3.36 − $13.44

The sale price is $13.44 and is greater than the cost of $12.

The sale price of the game is $13.44, and it is still greater than the cost of the game.

PRACTICE PROBLEMS

11.23 A book that has a selling price of $39.98 is on sale for 30% off. What is the sale price of the book?

11.24 A wholesaler sells a tent for a list price of $250. The wholesaler offers the retailer a trade discount of 35%. The retailer sells the tent for a 45% markup on the cost of the tent. At the end of the summer, the retailer puts the tent on sale with a 15% discount. Before Christmas, the retailer takes an additional 40% off the sale price. What is the final sale price of the tent? Does the sale price cover the cost the retailer paid for the tent from the wholesaler?

Chapter

12

Inventory

*I*nventory, or stock, is a business's merchandise, raw materials, and finished or unfinished products that have not been sold. Businesses must order, track, and value their inventory.

12.1 Ordering Inventory

There are many costs associated with ordering inventory, such as the wages for the personnel who generate a purchase order, office costs, and overhead costs. Office costs include the costs of forms, envelopes, stamps, telephones, faxes, copies, and computers. Overhead costs include things like utilities and maintenance. In addition, there are costs associated with handling the stock when it arrives; personnel must check the stock order and store the stock.

A business may find the average monthly cost of ordering stock, or calculate the average cost of a single purchase order. These costs do not include the cost of purchasing the stock, only the expenses that are incurred to generate the order and handle the stock.

Example: The PowerFlame Company issues about 150 purchase orders per month. The estimated monthly costs are as follows: 35% of the company's office personnel monthly cost of $8,500, $225 per hour for 10 hours for handling arriving stock, 15% of the overhead costs of $5,000. What is the average monthly cost of purchase orders?

Solution: Find the monthly cost for each expense by multiplying each expense by the percent or the number of hours. Add all of the expenses to find the total cost of purchasing for the month.

Personnel costs = $8,500 × 35% = $8,500 × 0.35 = $2,975

Handling costs = $225 × 10 = $2,250

Overhead costs = $5,000 × 15% = $5,000 × 0.15 = $750

Monthly cost of purchase orders = $2,975 + $2,250 + $750 = $5,975

Average cost per purchase order = $5,975 ÷ 150 = $39.83

The total monthly cost for purchasing stock is $5,975, and the average cost per purchase order is $39.83.

PRACTICE PROBLEMS

12.1 A department store issues 5,000 purchase orders per year. Three employees spend 35% of their time ordering stock. Their total wages and compensation are $130,000. The warehouse allocates 15% of their total expenses to ordering and handling stock, and 25% of the office costs are charges to ordering. If the total warehouse expenses per year are $275,000 and the total office costs are $125,000, what is the total yearly expense allocated to purchasing?

12.2 From the previous problem, what is the average cost per purchase order, to the nearest cent?

12.2 Tracking Inventory

A business must track its inventory. Stock that is received must be added to a stock record, and stock that is issued, or sold, must be deducted from the stock record. Tracking stock is much like keeping track of a checking account balance. A stock record might look like the one shown in Table 12.1.

Table 12.1 Sample Stock Record

Item: Queen Mattress Stock #PSN-1539		Reorder Point: 45 Unit: Each	
Date	*Quantity Received*	*Quantity Issued*	*Balance*
10/1			78
10/2		22	56
10/3		20	36
10/6	51		87
10/9		10	77

Example: Use the stock record from Table 12.1. If 40 mattresses were sold on 10/11, and 75 mattresses were received on 10/15, what is the balance on 10/15?

Solution: To the balance on 10/9, subtract the number of mattresses that were issued and add the number of mattresses that were received.

Balance on 10/11 = 77 − 40 = 37

Balance on 10/15 − 37 + 75 = 112

The balance on 10/15 is 112 mattresses.

PRACTICE PROBLEMS

12.3 An electronics wholesaler stocks several types of MP-3 players. The transactions for one MP-3 player are as follows: balance on 9/1, 165; sold 80 on 9/3; received 150 on 9/6; sold 50 on 9/8; sold 85 on 9/10; received 200 on 9/15; sold 125 on 9/18. What is the balance of the inventory on 9/18?

12.4   A toy manufacturer sells the following amounts of a toy to retailers
in November: 11/2, 150; 11/3, 580; 11/6, 1,250; 11/8, 487. The
manufacturer produced and placed the following amounts of
the toy in the inventory: 11/3, 250; 11/10, 1,000; 11/17, 1,000.
If the beginning balance for the toy on 11/1 was 1,800, what is
the balance on 11/17?

12.3 Reordering Inventory

As items in its inventory runs low, a business must reorder stock. For each
item that they stock, a business will identify a *reorder point*, which is the
minimum level of inventory that is allowed. A business must estimate
how quickly it sells or uses the item and the *lead time*, the amount of time
it takes to receive the item once it is ordered. To allow time for delays, a
business will often build in a *margin*, or an amount of safety stock, so that
it will not run out before an order is received.

There are several formulas that can be used to calculate the reorder point.
One formula uses the daily usage, lead time, and the safety stock:

> Reorder Point = (Daily Usage × Lead Time) + Safety Stock

Example: On average, a wholesaler sells 3,285 area rugs in 15 working
days. The lead time for ordering new stock is three working days from
the day the order is received. If the wholesaler wants to keep a safety
stock of 75 rugs, what is the reorder point for the area rug?

Solution: Divide the number of area rugs sold by the number of days to
find the average number sold per day. Multiply by the lead time and add
the safety stock to find the reorder point.

Daily usage = 3,285 rugs ÷ 15 days = 219 rugs per day

Reorder point = (219 × 3) + 75 = 657 + 75 = 732

The rugs must be reordered by the time the stock reaches 732 rugs.

PRACTICE PROBLEMS

12.5   A retailer sells 150 cans of primer in 30 working days. The lead
time is four working days, and the retailer wants to keep a safety
stock of 15 cans of primer. What is the reorder point?

12.6 A lawnmower manufacturer uses 5,000 belts every 25 working days to produce its products. The lead time for the belts is six working days. If the safety stock is three days' stock, what is the reorder point?

12.4 Valuing Inventory

Businesses must calculate the value of their inventory at different times for tax and insurance purposes and for financial statements and records. The value of the inventory is calculated by multiplying each stock item by its cost. While this may seem to be a simple calculation, it is made more complex by the fact that the cost of a certain stock item may vary so that the cost per item is different when stock is reordered.

Table 12.2 shows an inventory record that displays when stock was received and the cost of the stock. Notice that on June 30 there were 150 cameras left in the inventory, and that the digital cameras in stock were purchased at different prices.

Table 12.2 Inventory Record

Digital Camera #PZ11041

Date	Units	Cost	Total Value
Beginning Inventory			
April 1	100	$30	$3,000
Purchases			
April 12	150	$28	$4,200
May 2	85	$33	$2,805
May 22	125	$29	$3,625
June 10	150	$27	$4,050
June 25	75	$26	$1,950
Ending Inventory			
June 30	150		???

There are several methods for valuing the stock that is left in inventory. The three most widely used methods of valuing inventory are First In, First Out (FIFO); Last In, First Out (LIFO); and Weighted Average.

First In, First Out

The *First In, First Out*, or *FIFO* method assumes that the goods that are purchased first are then issued or sold first. So the cost of the inventory item is based on the most recent purchases. This method is widely used because the value of the inventory should be the closest to the cost of actually replacing the items in the inventory.

Using the inventory record in Table 12.2, the 150 digital cameras that remain in inventory are valued from the last sets of cameras that were purchased.

Example: What is the value of the 150 digital cameras left in inventory from Table 12.2 using the FIFO method?

Solution: Starting with the last purchase and moving backward, find the value for 150 of the cameras at the cost on that date. Multiply the number of units by the cost and add the costs together.

June 25: 75 cameras $\times$ $26 = $1,950

Cameras remaining to be valued $= 150 - 75 = 75$

June 10: 75 cameras $\times$ $27 = $2,025

Total value of 150 cameras $= $1,950 + $2,025 = $3,975

Using the FIFO method, the value of the 150 cameras is $3,975.

Last In, First Out

The *Last In, First Out*, or *LIFO* method assumes that the goods that are purchased last are sold first.

So the cost of the inventory item is based on the goods purchased first. The goods are not necessarily being sold in that order, but the inventory is valued using the costs of the goods purchased first.

Using the inventory record in Table 12.2 the 150 digital cameras that remain in inventory are valued from the beginning inventory and the first set of cameras purchased on April 12.

Example: What is the value of the 150 digital cameras left in inventory from Table 12.2 using the LIFO method?

Solution: Starting with the beginning inventory and the first purchases, find the quantities and the cost of 150 units. Multiply the number of units by the cost and add the costs together.

Beginning inventory: 100 cameras $\times$ \$30 = \$3,000

Cameras remaining to be valued = 150 − 100 = 50

April 12: 50 cameras $\times$ \$28 = \$1,400

Total value of 150 cameras = \$3,000 + \$1,400 = \$4,400

Using the LIFO method, the value of the 150 cameras is \$4,400.

Weighted Average

The *weighted average* method values the inventory at the average cost of the beginning inventory plus the cost of the purchases during the time period.

Example: What is the value of the 150 digital cameras left in inventory from Table 12.2 using the weighted average method?

Solution: Find the total cost of the beginning inventory and the inventory purchased. Find the total number of units in the beginning inventory and the purchases. Divide the total cost by the total number of units to find the average cost. Multiply the ending inventory by the average cost.

Beginning inventory: 100 $\times$ \$30 = \$3,000

4/12: 150 $\times$ \$28 = \$4,200

5/2: 85 $\times$ \$33 = \$2,805

5/22: 125 $\times$ \$29 = \$3,625

6/10: 150 $\times$ \$27 = \$4,050

6/25: 75 $\times$ \$26 = \$1,950

Total cost = \$19,630

Total number of units = 100 + 150 + 85 + 125 + 150 + 75 = 685

Average cost per unit = \$19,630 $\div$ 685 = \$28.66

Value of 150 units = 150 $\times$ \$28.66 = \$4,299

Using the weighted average method, the value of the 150 digital cameras is \$4,299.

PRACTICE PROBLEMS

12.7   A wholesaler that stocks basketballs has a beginning balance on October 1 of 450 basketballs with a cost of $7 each. The following stock purchases were made and added to the inventory: 10/15, 125 @ $7.50; 10/31, 200 @ $8.25; 11/14, 250 @ $9; 12/15, 75 @ $10.50; 12/22, 50 @ $11. The inventory on December 31 is 175 basketballs. What is the value of the inventory on December 31 using the FIFO method?

12.8   From the previous exercise, what is the value of the inventory on December 31 using the LIFO method?

12.9   What is the value of the inventory on December 31 using the weighted average method?

12.5 Carrying Inventory

The *carrying costs* of inventory include all of the costs of keeping the inventory until it is sold. Typically, carrying costs are expressed as a total amount per year or the cost to hold one unit of inventory for a single year. Carrying costs include interest on the money borrowed to purchase the inventory, personal property taxes on the inventory's value, insurance, storage costs, and loss.

Example: Music to My Ears Music Shop has an inventory of 25,000 music CDs valued at $14.95 per CD. The store pays 8% interest and property tax of 2.5% of the total inventory before losses. Each year, 5% of their CD inventory is either damaged or stolen. The other costs to carry the inventory total $9,500. What is the average carrying cost per CD?

Solution: Multiply the number of CDs by the value of each to find the total value of the inventory. Multiply the total value by the interest rate to find the amount of yearly interest. Multiply the inventory value by 2.5% to find the property tax and by 5% to find the amount of loss. Add the interest, property tax, loss and other costs to find the total annual carrying costs. Divide the annual carrying costs by the number of CDs to find the average cost per CD.

Value of inventory = 25,000 × \$14.95 = \$373,750

Interest = \$373,750 × 8% = \$373,750 × 0.08 = \$29,900

Property tax = \$373,750 × 2.5% = \$373,750 × 0.025 = \$9,343.75

Loss = \$373,750 × 5% = \$373,750 × 0.05 = \$18,687.50

Total carrying costs = \$29,900 + \$9,343.75 + \$18,687.50 + \$9,500 = \$67,431.25

Carrying cost per CD = \$67,431.25 ÷ 25,000 = \$2.70

The average annual carrying cost per CD is \$2.70.

PRACTICE PROBLEMS

12.10 An electronics wholesaler carries an inventory valued at \$550,000. The inventory carries interest payments of 9%, taxes of 5%, and insurance at 2% of the inventory's total value before losses. Loss is estimated at 8% of the value. Additional carrying costs are \$45,000. What are the annual carrying costs for the inventory?

12.11 Instead of calculating carrying costs, some businesses use an estimate. A standard rule of thumb is to estimate carrying costs at 25% of the inventory's value. Using this standard, what is the annual carrying cost for the electronics wholesaler in the previous exercise?

Chapter

13

Financial Statements and Ratios

13.1 Income Statement

13.2 Balance Sheet

13.3 Financial Ratios

Businesses create quarterly or yearly financial statements to show the business's financial position. The two basic types of financial statements are the income statement and the balance sheet. The information contained in these financial statements can be used to calculate financial ratios that give a snapshot of different financial aspects of the business.

13.1 Income Statement

An *income statement* shows the net income of a business over a period of time, such as a month, quarter, or year. If there is more revenue than expenses, there is a profit. If there are more expenses than revenue, there is a loss.

The five major sections of an income statement include revenue, cost of goods sold, gross profit, operating expenses, and net income. A sample income statement for Unique Expressions Gift Shop is shows in Table 13.1.

Table 13.1 Sample Income Statement

For the month ended August 31, 20—

Revenue		
Sales	151,904	
Less sales returns and allowances	3,800	
Net Sales		148,104
Cost of Goods Sold		
Beginning inventory August 1	432,276	
Purchases	63,096	
Goods available for sale	495,372	
Ending inventory, August 31	402,840	
Cost of goods sold		92,532
Gross profit on sales		55,572
Operating expenses		
Salaries and wages	21,134	
Rent	6,388	
Taxes	3,728	
Utilities	3,356	
Advertising	1,560	
Depreciation of equipment	1,452	
Insurance	506	
Other expenses	1,708	
Total operating expenses		39,832
Net Income		15,740

Net Sales

Sales are reduced by returns and allowances, or reductions in price offered to customers for damaged goods. To find *net sales*, subtract returns and allowances from the sales.

Example: The annual sales for Stonegate Industries were $975,000. Returns and allowances for the year totaled $2,500. What were the net sales for the year?

Solution: Subtract the returns and allowances from the sales.

Net sales = $975,000 − $2,500 = $972,500

The net sales for the year were $972,500.

PRACTICE PROBLEM

13.1 For the month of January, Jackson's Lumber and Supply had sales of $200,000. For the same period, returns and allowances totaled $2,675. What are the net sales for January?

Cost of Goods Sold

The *cost of goods sold* are the costs directly associated with buying or producing the good sold. For a retailer, the cost of goods sold is the cost of purchasing the merchandise. For a manufacturer, the cost of goods sold is the prime cost to manufacture the goods. The prime cost is the sum of the cost of the raw materials and the direct labor costs required to manufacture the goods.

The cost of goods sold is calculated using inventory records and valuation. The beginning inventory plus the stock added through purchase or manufacture shows the value of the goods available for sale during the time period of the income statement. Subtracting the ending inventory gives the cost of the goods that were sold.

Cost of Goods Sold = Beginning Inventory + Cost of Goods Purchased or Produced − Ending Inventory

Example: The Fashion Place has an inventory valued at $875,000 on January 1. During January, stock costing $235,800 was purchased. At the end of January, the merchandise inventory is $685,255. What is the cost of goods sold for January?

Solution: Add the cost of goods purchased to the beginning inventory, and subtract the ending inventory.

Cost of goods sold = $875,000 + $235,800 − $685,255 = $425,545

The cost of goods sold for January is $425,545.

PRACTICE PROBLEMS

13.2   Johnson's Paint Store's beginning inventory in April was valued at $527,836. During April, goods costing $302,400 were purchased. At the end of the month, the inventory was valued at $638,525. What is the cost of goods sold for April?

13.3   Rivers Manufacturing Company has a beginning inventory of $1,847,500 on January 1. The cost of goods manufactured during the year was $2,834,765, and the ending inventory on December 31 was $1,655,150. What is the cost of goods sold for the year?

Gross Profit

The *gross profit* is the difference between the net sales and the cost of goods sold. Gross profit does not take into consideration the other costs of doing business.

Gross Profit = Net Sales − Cost of Goods Sold

Example: The Luggage Emporium had net sales of $87,657 in October. The cost of goods sold in October was $43,775. What was the gross profit for October?

Solution: Subtract the cost of goods sold from the net sales.

Gross profit = $87,657 − $43,775 = $43,882

The gross profit for October was $43,882.

PRACTICE PROBLEMS

13.4 Jerseys and More had net sales last year of $1,487,950. During the same period, the cost of goods sold was $338,460. What was the gross profit last year?

13.5 Simpson Manufacturing had a beginning inventory valued at $653,828. During the quarter, goods costing $250,595 were produced. The inventory at the end of the quarter was $515,787. During the quarter, the sales totaled $985,462 while returns and allowances were $18,500. What is the gross profit for the quarter?

Operating Expenses and Net Income

Operating expenses are ongoing expenses for running a business. Operating expenses include things like salaries and wages, taxes, utilities, rent, advertising, depreciation, and office expenses. Specific operating expenses will vary by business. For a manufacturing business, the prime costs that are accounted for in the cost of goods sold are not included as operating expenses.

On an income statement, the operating expenses are added to find the total operating expenses for the time period of the income statement.

The *net income* or net loss is found by subtracting the total operating expenses from the gross profit. The net income or loss is the profit or loss realized after accounting for the expenses.

$$\text{Net Income} = \text{Gross Profit} - \text{Total Operating Expenses}$$

If the total operating expenses are greater than the gross profit, the business has a *net loss*. Typically a net loss will be recorded on an income sheet in parentheses. For example, a net loss of $10,000 will be shown as ($10,000).

Example: Ned's Outdoor Sports had a gross profit of $687,500 last year, and total expenses of $690,432. What is the store's net income or loss?

Solution: Subtract the total operating expenses from the gross profit.

Net income or loss = $687,500 − $690,432 = −$2,932 or ($2,932)

The store had a net loss of $2,932.

Practice Problems

13.6 Mailboxes Galore had a gross profit of $42,876 in March. The expenses for the month included: salaries and wages, $12,500; rent, $1,500; taxes, $720; utilities, $580; advertising, $225; depreciation, $310; insurance, $142; and other expenses, $512. What was the store's net income or loss?

13.7 Prepare an income statement for Gore Manufacturing for the month of March using the following information: sales, $215,000; returns and allowances, $9,026; beginning inventory $23,179; cost of goods sold $158,107; ending inventory, $20,486; salaries, $19,487; office expenses, $3,256; payroll taxes, $2,612; and depreciation, $2,788.

13.2 Balance Sheet

A *balance sheet* shows a company's financial status at a single point in time. Balance sheets are prepared by businesses at least once a year. The three sections of a balance sheet are assets, liabilities, and equity.

Assets are things a business owns that have value, such as cash, accounts receivable, inventory, property, equipment, and buildings.

Accounts receivable are customer's credit accounts. *Current assets* are assets that can be turned into cash or used within a year, such as cash and accounts receivable. *Long-term assets* are assets that have a useful life of more than a year such as equipment, land, and buildings.

Liabilities are the debts and other financial obligations that a business has.

Accounts payable and loans are examples of liabilities. *Accounts payable* are the credit accounts that the business owes to another business, such as vendors, lenders, and bond holders.

A business will often buy inventory or raw materials on credit from a supplier. *Current liabilities* are debts that are due in a short time, such as accounts payable, while *long-term liabilities* are longer term debts, such as a bank loan.

Equity represents the owner's share in a business. If there are no liabilities, then the assets are equal to the equity. If there are liabilities, then the equity is reduced by the amount of liability.

$$\text{Equity} = \text{Assets} - \text{Liabilities}$$

A sample balance sheet for the Unique Expressions Gift Shop is shown in Table 13.2.

Table 13.2 Sample Balance Sheet

August 31, 20—		
Assets		
Cash	$ 32,600	
Merchandise Inventory	393,840	
Store Supplies	10,760	
Store Equipment	49,000	
Total Assets		$486,200
Liabilities		
Accounts Payable	$103,120	
Bank Loan	90,400	
Total Liabilities		$193,520
Equity		
Sharon Jones, Equity		292,680
Total Liabilities and Equity		$486,200

Example: One year later, Unique Expressions Gift Shop has assets worth $640,870 and liabilities worth $155,493. What is the owner's equity?

Solution: Subtract the liabilities from the assets.

Equity = $640,870 − $155,493 = $485,377

Sharon Jones has $485,377 of equity in the business.

PRACTICE PROBLEMS

13.8 The Shoe Market has these assets on July 1: cash, $15,842.92; accounts receivable, $62,946.33; inventory, $539,429.22; supplies, $6,482; equipment, $42,156; and building, $175,000. The company owes $285,936 in accounts payable and $85,000 for a bank loan. What is the owner's equity in the business?

13.9 On March 1, Jones and Sons has the following assets: cash, $7,260, accounts receivable, $15,492; equipment, $35,844; and supplies, $1,225. The business owes $14,497 in accounts payable and $9,500 for a bank loan. What is the owner's equity in the business?

13.3 Financial Ratios

Financial ratios can help business owners, shareholders, and prospective investors analyze the information contained in the income statement and balance sheet. Ratios can be compared over time to evaluate changes in a business's financial condition.

In addition, many industries calculate average industry ratios so that businesses can compare their performance to others in the same industry.

Profit Margin

Profit margin is an indicator of a business's profitability, or the business's ability to generate a return on its resources. *Gross profit margin* compares the gross profit to net sales and shows what percent of every dollar used to purchase raw materials or inventory is turned into gross profit. Gross profit margin illustrates how efficient the business is in managing inventory, pricing, and production.

$$\text{Gross Profit Margin} = \frac{\text{Gross Profit}}{\text{Net Sales}}$$

Example: The Unique Expressions Gift Shop income statement in Table 13.1 shows net sales of $148,104 and gross profit of $55,572. What is the gross profit margin?

Solution: Divide the gross profit by the net sales and change the decimal to a percent.

$$\text{Gross Profit Margin} = \frac{\$55,572}{\$148,104} = 0.375 = 37.5\%$$

The gross profit margin is 37.5%.

Net profit margin compares the net income to net sales and shows what percent of every sales dollar is turned into profit. This ratio indicates how well a business is managing its operating expenses.

$$\text{Net Profit Margin} = \frac{\text{Net Income}}{\text{Net Sales}}$$

Example: The Unique Expressions Gift Shop income statement in Table 13.1 shows net sales of $148,104 and net income of $15,740. What is the net profit margin?

Solution: Divide the net income by the net sales and change the decimal to a percent.

$$\text{Net profit margin} = \frac{\$15,740}{\$148,104} = 0.106 = 10.6\%$$

The net profit margin is 10.6%

PRACTICE PROBLEMS

13.10 For the month of March, Gore Manufacturing had net sales of $205,974, gross profit of $45,174, and net income of $17,031. What is the gross profit margin to the nearest tenth of a percent?

13.11 What is the net profit margin to the nearest tenth of a percent for Gore Manufacturing from the previous exercise?

Merchandise Turnover Rate

The *merchandise turnover rate* is the number of times per period that a business sells its average stock of merchandise. Although different businesses have different turnover rates, an especially high or low turnover rate as compared to other businesses in the same industry can indicate over-stocking, under-stocking, or obsolete merchandise.

To find the merchandise turnover rate, divide the cost of goods sold for the period by the average merchandise inventory for the period. The average merchandise inventory for the period is the average of the beginning and ending inventory for the period.

- $$\text{Merchandise Turnover Rate} = \frac{\text{Cost of Goods Sold for Period}}{\text{Average Merchandise Inventory for Period}}$$

- $$\text{Average Merchandise Inventory} = \frac{\text{Beginning Inventory} + \text{Ending Inventory}}{2}$$

Example: The Unique Expressions Gift Shop income statement in Table 13.1 shows a beginning inventory of $432,276, an ending inventory of $402,840, and the cost of goods sold as $92,532. What is the merchandise turnover rate for this period?

Solution: Add the beginning and ending inventory and divide by 2 to find the average merchandise inventory for the period. Divide the cost of goods sold by the average.

$$\text{Average merchandise inventory} = \frac{\$432,276+\$402,840}{2} = \frac{\$835,116}{2} = \$417,558$$

$$\text{Merchandise turnover rate} = \frac{\$92,532}{\$417,558} = 0.22$$

The Unique Expressions Gift Shop turned over 0.22 of its merchandise during the period.

A merchandise turnover rate of less than 1 shows that the entire inventory was not sold during the period, while a merchandise inventory turnover rate of greater than 1 shows how many times the entire inventory was sold during that time. The Unique Expressions Gift Shop sold less than one quarter of its inventory in the month of August.

PRACTICE PROBLEMS

13.12 On January 1, The Jewelry Palace had a beginning inventory of $1,863,225. On December 31, the ending inventory was $1,563,480. The cost of goods sold for the year was $30,840,350. What was the merchandise turnover rate to the nearest tenth for the year?

13.13 On January 1, a manufacturer had an inventory stock item costing $68,000. On March 31, the inventory was $72,585. The cost of goods sold for the item for the quarter was $630,000. What was the turnover rate to the nearest tenth for the stock item for the quarter?

Current Ratio

The *current ratio* compares current assets to current liabilities. Recall that current assets are assets that can be turned into cash or used within a year, while current liabilities are short-term debts, such as accounts payable. The current ratio is a measure of a business's ability to pay current liabilities with current assets. A current ratio of 2 to 1 may be expressed as 2 to 1, 2:1, or just 2.

$$\text{Current Ratio} = \frac{\text{Current Assets}}{\text{Current Liabilities}}$$

Example: Use Unique Expressions Gift Shop's balance sheet in Table 13.2 to find the current ratio for the shop.

Solution: Identify the current assets and find the sum. Identify the current liabilities and find the sum. Divide the current assets by the current liabilities.

Current assets = $32,600 + $393,840 + $10,760 = $437,200

Current liabilities = $103,120

$$\text{Current Ratio} = \frac{\$437,200}{\$103,120} = 4.2$$

The current ratio for Unique Expressions Gift Shop is 4.2:1

The Unique Expressions Gift Shop has more than four times more current assets than liabilities.

PRACTICE PROBLEMS

13.14 What is the current ratio to the nearest tenth for the Shoe Market in Exercise 13.8?

13.15 What is the current ratio to the nearest tenth for Jones and Sons in Exercise 13.9?

Debt-to-Equity Ratio

The *debt-to-equity ratio* compares the amount of debt, or liabilities, to the equity, or net worth, of the business. A higher ratio indicates a greater financial risk to current or future creditors or investors because there is a greater likelihood that the business will not generate enough cash to pay their financial obligations when due. The debt-to-equity ratio is usually expressed as a percent.

$$\text{Debt-to-Equity Ratio} = \frac{\text{Total Liabilities}}{\text{Equity}}$$

Example: Use Unique Expressions Gift Shop's balance sheet in Table 13.2 to find the debt-to-equity ratio for the shop.

Solution: Divide the total liabilities by the owner's equity and change the decimal into a percent.

$$\text{Debt-to-equity ratio} = \frac{\$193,520}{\$292,680} = 0.661 = 66.1\%$$

PRACTICE PROBLEMS

13.16 What is the debt-to-equity ratio to the nearest tenth for the Shoe Market in Exercise 13.8?

13.17 What is the debt-to-equity ratio to the nearest tenth for Jones and Sons in Exercise 13.9?

Return on Equity

The *return on equity*, also called return on investment, compares the net income to the equity. It shows the rate of return on the money invested in the business. The return on equity is found by dividing the net income from the income statement by the equity shown on the balance sheet. Return on equity is typically expressed as a percent.

$$\text{Return on Equity} = \frac{\text{Net Income}}{\text{Equity}}$$

Example: Using Tables 13.1 and 13.2, find the return on equity for the Unique Expressions Gift Shop.

Solution: Identify the net income and the equity from the income statement and balance sheet. Divide the net income by the equity and change the decimal to a percent.

$$\text{Return on equity} = \frac{\$15,740}{\$292,680} = 0.054 = 5.4\%$$

The return on equity is 5.4%.

PRACTICE PROBLEMS

13.18 T&J Hardware had a net income of $125,600 on the end-of-the-year income statement. The balance sheet on the same day showed equity of $571,000. What is the return on equity to the nearest tenth of a percent?

13.19 The Golf Shop's December 31 income statement showed the following: net sales $482,955; cost of goods sold, $257,800; and total operating expenses, $122,729. The balance sheet showed equity of $355,823. What was the return on equity to the nearest tenth of a percent?

Financial statements and ratios can help business owners to evaluate the current status of the business, to set goals, and to make financial decisions that will benefit the business.

Appendix

Answers to Practice Problems

Chapter 1

1.1 $\dfrac{2 \times 4}{3 \times 4} = \dfrac{8}{12}$

1.2 $\dfrac{10 \div 2}{12 \div 2} = \dfrac{5}{6}$

1.3 $\dfrac{4 \div 4}{12 \div 4} = \dfrac{1}{3}$

1.4 $\dfrac{18 \div 6}{42 \div 6} = \dfrac{3}{7}$

1.5 $\dfrac{25}{4} = 6\dfrac{1}{4}$

1.6 $5\dfrac{7}{8} = \dfrac{8 \times 5 + 7}{8} = \dfrac{47}{8}$

1.7 $3\dfrac{1}{4} = 3\dfrac{1 \times 3}{4 \times 3} = 3\dfrac{3}{12}$

$2\dfrac{2}{3} = 2\dfrac{2 \times 4}{3 \times 4} = 2\dfrac{8}{12}$

$3\dfrac{3}{12} + 2\dfrac{8}{12} = 5\dfrac{3+8}{12} = 5\dfrac{11}{12}$

1.8 $8\dfrac{6}{7} = 8\dfrac{6 \times 2}{7 \times 2} = 8\dfrac{12}{14}$

$8\dfrac{12}{14} + 1\dfrac{5}{14} = 9\dfrac{12+5}{14} = 9\dfrac{17}{14}; \ 9\dfrac{17}{14} = 9 + 1\dfrac{3}{14} = 10\dfrac{3}{14}$

1.9 $3\dfrac{5}{8} = 3\dfrac{5 \times 3}{8 \times 3} = 3\dfrac{15}{24}$

$2\dfrac{1}{6} = 2\dfrac{1 \times 4}{6 \times 4} = 2\dfrac{4}{24}$

$3\dfrac{15}{24} - 2\dfrac{4}{24} = 1\dfrac{15-4}{24} = 1\dfrac{11}{24}$

1.10 $6\dfrac{1}{3} = 6\dfrac{1 \times 4}{3 \times 4} = 6\dfrac{4}{12}$

$6\dfrac{4}{12} - 5\dfrac{11}{12} = 5\dfrac{4+12}{12} - 5\dfrac{11}{12} = 5\dfrac{16}{12} - 5\dfrac{11}{12} = \dfrac{5}{12}$

1.11 $\dfrac{4}{5} \times \dfrac{2}{3} = \dfrac{4 \times 2}{5 \times 3} = \dfrac{8}{15}$

1.12 $\dfrac{3}{8} \div \dfrac{9}{16} = \dfrac{3}{8} \times \dfrac{16}{9} = \dfrac{3 \times 16}{8 \times 9} = \dfrac{48}{72} = \dfrac{48 \div 24}{72 \div 24} = \dfrac{2}{3}$

1.13 $2\dfrac{3}{8} \div \dfrac{4}{5} = \dfrac{8 \times 2 + 3}{8} \div \dfrac{4}{5} = \dfrac{19}{8} \times \dfrac{5}{4} = \dfrac{19 \times 5}{8 \times 4} = \dfrac{95}{32} = 2\dfrac{31}{32}$

1.14 829,500

1.15 4,389.30

1.16 3.517

1.17 11.85

1.18 0.0111

1.19 52.1

1.20 $2 \div 5 = 0.4$

1.21 $4 \div 11 = 0.36$

1.22 $0.475 = \dfrac{475}{1,000} = \dfrac{475 \div 25}{1,000 \div 25} = \dfrac{19}{40}$

1.23 $23\% = \dfrac{23}{100} = 0.23$

1.24 $120\% = \dfrac{120}{100} = 1.2$

1.25 $0.35 = 35\%$

1.26 $\dfrac{4}{5} = 0.8 = 0.80 = 80\%$

1.27 $28\% = \dfrac{28}{100} = \dfrac{28 \div 4}{100 \div 4} = \dfrac{7}{25}$

1.28 $75 \times 22\% = 75 \times 0.22 = 16.5$

1.29 $125 \times 4\% = 125 \times 0.04 = 5$

1.30 $\dfrac{16}{64} = 0.25 = 25\%$

1.31 $\dfrac{22}{140} = 0.157 = 15.7\%$

1.32 $G = 45 \times \$10.25 = \$461.25$

1.33 $I = \$2{,}500 \times 2\% \times 3 = \$2{,}500 \times 0.02 \times 3 = \$150$

1.34 $A = 1{,}500(1 + 0.015)^6 = 1{,}500(1.015)^6 = 1{,}640.16$

1.35 $APY = (1 + 0.0125)^{10} - 1 = (1.0125)^{10} - 1 = 1.13 - 1 = 0.13$

Chapter 2

2.1 $8 \times 5 = 40$ hours

$40 \times \$16.85 = \$674$

2.2 $9\dfrac{1}{2} + 5 + 8\dfrac{3}{4} + 10 = 33\dfrac{1}{4} = 33.25$ hours;

$33.25 \times \$6.75 = \$224.44$

2.3 $\$25.50 \times 1.5 = \$38.25$

$11\dfrac{1}{2} - 8 = 3\dfrac{1}{2}; \; 8 \times \$25.50 = \$204;$

$3\dfrac{1}{2} \times \$38.25 = \$133.88; \; \$204 + \$133.88 = \$337.88$

2.4 $\$10.75 \times 1.5 = \$16.125$

$\$10.75 \times 2 = \$21.50$

$40 \times \$10.75 = \$430$

$5 \times \$16.125 = \$80.63$

$5 \times \$21.50 = \$107.50$

$\$430 + \$80.63 + \$107.50 = \$618.13$

2.5 $2,000 \div 2 = 1,000$

2.6 $36,000 \div 12 = 3,000$

No, the gross pay is less than $3,250 per month.

2.7 $950 \times 52 = 49,400$

The job with weekly pay has a higher pay rate.

2.8 $50 \times 8.50 = 425$

2.9 $325,000 \times 3\% = 325,000 \times 0.03 = 9,750$

2.10 $55,000 - 10,000 = 45,000$

$45,000 \times 2\% = 45,000 \times 0.02 = 900$

$900 + 1,200 = 2,100$

2.11 $125,000 - 75,000 = 50,000$

$50,000 \times 10\% = 50,000 \times 0.10 = 5,000$

$40 \times 6.50 = 260$

$260 + 5,000 = 5,260$

2.12 $100 - 30 = 70$

$30 \times 2.50 = 75$

$70 \times 3.25 = 227.50$

$75 + 227.50 = 302.50$

2.13 $20,000 \times 4\% = 20,000 \times 0.04 = 800$

$20,000 \times 8\% = 20,000 \times 0.08 = 1,600$

$150,000 - 40,000 = 110,000$

$110,000 \times 10\% = 110,000 \times 0.10 = 11,000$

$800 + 1,600 + 11,000 = 13,400$

2.14 $\dfrac{\$1.12}{\$8} = 0.14 = 14\%$

2.15 $150 \times \$9.75 = \$1,462.50$

$\$1,712.50 - \$1,462.50 = \$250$

$\dfrac{\$250}{\$5,000} = 0.05 = 5\%$

2.16 $400 \times \$4.50 = \$1,800$

2.17 $12 \times \$80 = \$960$

2.18 $4 \times 2 = 8$

$8 \times \$2 = \$16$

2.19 $\$85 \times 15\% = \$85 \times 0.15 = \$12.75$

2.20 $\$85 + \$48 + \$75 + \$125 + \$63 + \$230 = \$626$

$\$626 \div 5 = \$125.20$

2.21 $\$2,500 \times 2 = \$5,000$

$\$4,000 \times 3 = \$12,000$

$\$3,000 \times 6 = \$18,000$

$\$5,000 + \$12,000 + \$18,000 + \$8,000 = \$43,000$

$\$43,000 \div 12 = \$3,583.33$

2.22 $\$5,500 \times 5 = \$27,500$

$\$27,500 - \$15,000 = \$12,500$

2.23 $100,000 \times 12 = \$1,200,000$

$85,000 \times 2 = \$170,000$

$125,000 \times 3 = \$375,000$

$90,000 \times 4 = \$360,000$

$80,000 \times 2 = \$160,000$

$170,000 + \$375,000 + \$360,000 + \$160,000 = \$1,065,000$

$1,200,000 - \$1,065,000 = \$135,000$

Chapter 3

3.1 $18

3.2 $71 - \$40 = \$31$

3.3 $10,800 \times 5.65\% = \$10,800 \times 0.0565 = \$610.20$

3.4 $10,800 \times 11 = \$118,800$

Social Security limit passed

$10,800 \times 1.45\% = \$10,800 \times 0.0145 = \$156.60$

3.5 $106,800 \times 4.2\% = \$106,800 \times 0.042 = \$4,485.60$

$106,800 \times 6.2\% = \$106,800 \times 0.062 = \$6,621.60$

2011 maximum $= \$4,485.60$

6.2% maximum $= \$6,621.60$

3.6 $53.96 + \$78 + \$12 + \$45 = \$188.96$

$955 - \$188.96 = \$766.04$

3.7   Federal withholding = $91

$2,645 × 5.65% = $2,645 × 0.0565 = $149.44

$91 + $149.94 + $35 + $150 = $425.44

$2,645 − $425.44 = $2,219.56

3.8 $65,405 − $2,500 = $62,905

$10,850 + $3,650 = $14,500

$62,905 − $14,500 = $48,405

3.9 $22,000 − $500 = $21,500

$5,800 + $3,650 = $9,450

$21,500 − $9,450 = $12,050

3.10   Tax due = $3,524

$3,524 − $3,289 = $235

Amount due = $235

3.11   Tax due = $3,336

$3,336 − $3,023 = $313

Amount due = $313

3.12 $14,866 × 3% = $14,866 × 0.03 = $445.98

3.13 $55,450 − $3,000 = $52,450

$52,450 × 4.8% = $52,450 × 0.048 = $2,517.60

3.14 $10,000 + $15,000 = $25,000

$36,300 − $25,000 = $11,300

$10,000 × 2.59% = $10,000 × 0.0259 = $259

$15,000 × 2.88% = $15,000 × 0.288 = $432

$11,300 × 3.36% = $11,300 × 0.0336 = $379.68

$259 + $432 + $379.68 = $1,070.68

3.15 $\$2,000 \times 4.24\% = \$2,000 \times 0.0424 = \$84.80$

3.16 $\$48,940 \times 25\% = \$48,940 \times 0.25 = \$12,235$
$\$48,940 + \$12,235 = \$61,175$

3.17 $\$6,000 + \$5,600 + \$7,800 = \$19,400$
$\$86,000 + \$19,400 = \$105,400$

3.18 $\$38,395 \times 15\% = \$38,395 \times 0.15 = \$5,759.25$
$\$38,395 + \$5,759.25 = \$44,154.25$
$\$1,250 + \$240 + \$350 = \$1,840$
$\$44,154.25 - \$1,840 = \$42,314.25$

3.19 $\$74,800 \times 28\% = \$74,800 \times 0.28 = \$20,944$
$\$74,800 + \$20,944 = \$95,744$
$\$95,744 - \$4,200 = \$91,544$
$\$73,500 \times 20\% = \$73,500 \times 0.20 = \$14,700$
$\$73,500 + \$14,700 = \$88,200$
$\$88,200 - \$2,000 = \$86,200$
$\$91,544 - \$86,200 = \$5,344$
Current job offers the greatest net job benefits by $5,344.

3.20 $\$1,980 - \$330 = \$1,650$
$$\frac{\$1,650}{\$1,980} = 0.83 = 83\%$$

3.21   Federal withholding = $87
$\$2,625 \times 5.65\% = \$2,625 \times 0.0565 = \$148.31$
$\$87 + \$148.31 = \$235.31$
$\$2,625 - \$235.31 = \$2,389.69$
$$\frac{\$2,389.69}{\$2,625} = 91\%$$

3.22   Current federal withholding = $146

$2,400 × 5.65% = $2,400 × 0.0565 = $135.60

$2,400 × 2% = $2,400 × 0.02 = $48

$146 + $135.60 + $48 = $329.60

$2,400 − $329.60 = $2,070.40

$2,400 × 7% = $2,400 × 0.07 = $168

$2,400 + $168 = $2,568

New federal withholding = $170

$2,568 × 5.65% = $2,568 × 0.0565 = $145.09

$2,568 × 2% = $2,568 × 0.02 = $51.36

$170 + $145.09 + $51.36 = $366.45

$2,568 − $366.45 = $2,201.55

Gross pay increase = $2,568 − $2,400 = $168

Net pay increase = $2,201.55 − $2,070.40 = $131.15

3.23 $\dfrac{\$131.15}{\$2,070.40} = 0.06 = 6\%$

3.24   Current federal withholding = $34

$2,685 × 5.65% = $2,685 × 0.0565 = $151.70

$34 + $151.70 = $185.70

$2,685 − $250 = $2,435

Federal withholding = $6

$2,435 × 5.65% = $2,435 × 0.0565 = $137.58

$6 + $137.58 = $143.58

$185.70 − $143.58 = $42.12

$42.12 × 12 = $505.44

$42.12 per month

$505.44 per year

Chapter 4

4.1 $\$650 \times 1.5\% \times \dfrac{1}{12} = \$650 \times 0.015 \times \dfrac{1}{12} = \$0.81$

4.2 $\$25,000 \times 5\% \times \dfrac{1}{4} = \$25,000 \times 0.05 \times \dfrac{1}{4} = \$312.50$

4.3 $\$1,500 \times 2.25\% \times \dfrac{1}{2} = \$1,500 \times 0.0225 \times \dfrac{1}{2} = \$16.88;$

$\$1,500 + \$16.88 = \$1,516.88$

$\$1,516.88 \times 2.25\% \times \dfrac{1}{2} = \$1,516.88 \times 0.0225 \times \dfrac{1}{2} = \$17.06;$

$\$1,516.88 + \$17.06 = \$1533.94$

4.4 $\$6,000 \times 3.5\% \times \dfrac{1}{4} = \$6,000 \times 0.035 \times \dfrac{1}{4} = \$52.50;$

$\$6,000 + \$52.50 = \$6,052.50$

$\$6,052.50 \times 3.5\% \times \dfrac{1}{4} = \$6,052.50 \times 0.035 \times \dfrac{1}{4} = \$52.96;$

$\$6,052.50 + \$52.96 = \$6,105.46$

$\$6,105.46 \times 3.5\% \times \dfrac{1}{4} = \$6,105.46 \times 0.035 \times \dfrac{1}{4} = \$53.42;$

$\$6,105.46 + \$53.42 = \$6,158.88$

$\$6,158.88 \times 3.5\% \times \dfrac{1}{4} = \$6,158.88 \times 0.035 \times \dfrac{1}{4} = \$53.89;$

$\$6,158.88 + \$53.89 = \$6,212.77$

$\$6,212.77 - \$6,000 = \$212.77$

4.5 $\dfrac{4\%}{4} = 1\%;$

$10 \times 4 = 40$

Multiplier $= 1.488864$

$\$2,000 \times 1.488864 = \$2,977.73$

4.6 $\dfrac{6\%}{4} = 1.5\%;$

$5 \times 4 = 20$

Multiplier $= 1.346855$

$\$3,500 \times 1.346855 = \$4,713.99$

$\$4,713.99 - \$3,500 = \$1,213.99$

Total in account $\$4,713.99$

Interest $\$1,213.99$

4.7 $\dfrac{5\%}{2} = 2.5\% = 0.025;$

$5 \times 2 = 10$

$A = \$6,500(1 + 0.025)^{10} = \$6,500(1.025)^{10} = \$8,320.55$

4.8 $\dfrac{4\%}{4} = 1\% = 0.01;$

$20 \times 4 = 80$

$A = \$15,000(1 + 0.01)^{80} = \$15,000(1.01)^{80} = \$33,250.73$

$\$33,250.73 - \$15,000 = \$18,250.73$

4.9 $\dfrac{3\%}{4} = 0.75\% = 0.0075;$

Number of periods $= 4$

$APY = (1 + 0.0075)^{4} - 1 = (1.0075)^{4} - 1 = 0.0303 = 3.03\%$

4.10 $\dfrac{2\%}{12} = 0.1667\% = 0.001667;$

Number of periods $= 12$

$\text{APY} = (1 + 0.001667)^{12} - 1 = (1.001667)^{12} - 1$
$\phantom{\text{APY}} = 0.0202 = 2.02\%$

4.11 $\$600 \times 1.5\% = \$600 \times 0.015 = \$9$

4.12 $\$1,000 \times 4.5\% = \$1,000 \times 0.045 = \$45$

$\$1,000 \times 3.2\% = \$1,000 \times 0.032 = \$32$

$\$45 - \$32 = \$13$

4.13 $\$1,000 \times 6\% \times 5 = \$1,000 \times 0.06 \times 5 = \$300$

4.14 $\$25,000 \times 5\% \times 1 = \$25,000 \times 0.05 \times 1 = \$1,250$

$\$1,250 \div 4 = \$312.50$

4.15 $\$4,000 \times 5\% \times \dfrac{1}{4} = \$4,000 \times 0.05 \times \dfrac{1}{4} = \$50$

4.16 $\$2,000 \times 3.5\% \times \dfrac{2}{12} = \$2,000 \times 0.035 \times \dfrac{2}{12} = \$11.67;$

$\$2,000 \times 3.5\% \times 1 = \$2,000 \times 0.035 \times 1 = \$70$

$\$2,000 + \$70 = \$2,070$

$\$2,070 - \$11.67 = \$2,058.33$

4.17 $\dfrac{3\%}{12} = 0.25\%;$

$2 \times 12 = 24$

Multiplier $= 24.70282$

$\$200 \times 24.70282 = \$4,940.56$

4.18 $\dfrac{8\%}{4} = 2\%;$

$4 \times 4 = 16$

Multiplier $= 18.63929$

$\$1,000 \times 18.63929 = \$18,639.29$

4.19 $\dfrac{8\%}{2} = 4\%;$

$10 \times 2 = 20$

Multiplier $= 13.59033$

$\$5,000 \times 13.59033 = \$67,951.65$

4.20 $\dfrac{2\%}{4} = \dfrac{1}{2}\%;$

$6 \times 4 = 24$

Multiplier $= 22.56287$

$\$200 \times 22.56287 = \$4,512.57$

4.21 $\$468.22 + \$75.43 + \$122.89 = \$666.54$

$\$666.54 - \$150 = \$516.54$

4.22 $\$50 \times 5 = \$250$

$\$20 \times 10 = \$200$

$\$5 \times 6 = \$30$

$\$250 + \$200 + \$30 + \$2.34 = \$482.34$

4.23 $\$122.31 + \$653.82 - \$25.69 - \$99.86 - \$43.82 = \$606.76$

4.24 $\$1,486.22 + \$435 - \$200 - \$305 - \$75.22 = \$1,341$

4.25 $\$11.43 + \$121.86 = \$133.29$

$\$589.22 + \$263 = \$852.22$

$\$231.85 + \$852.22 - \$133.29 = \$950.78$

4.26 $45.83 + $75 + $82.19 = $203.02

$525 + $84.39 = $609.39

$984.39 + $609.39 − $203.02 = $1,390.76

Chapter 5

5.1 $2,480 × 3% = $2,480 × 0.03 = $74.40

$2,480 + $74.40 + 10 = $2,564.40

5.2 $3,400 × 5% = $3,400 × 0.05 = $170

$68.43 × 12 = $821.16

$170 + $821.16 + $75 = $1,066.16

5.3 18% ÷ 12 = 1.5% = 0.015

$4,329.18 × 0.015 = $64.94

5.4 15% ÷ 365 = 0.0411% = 0.000411

$822.49 × 0.000411 × 30 = $10.14

5.5 20% ÷ 12 = 1.6667% = 0.016667

$4,821.13 × 0.016667 = $80.35

$4,821.13 + ($80.35 + $529.74) − $1,485 = $3,946.22

Finance charge = $80.35

New balance = $3,946.22

5.6 12% ÷ 365 = 0.0329% = 0.000329

$692.28 × 0.000329 × 30 = $6.83

$692.28 + ($6.83 + $238.92) − $70 = $868.03

Finance charge = $6.83

New balance = $868.03

5.7 $20\% \div 12 = 1.6667\% = 0.016667$

$\$4,821.13 - \$1,485 = \$3,336.13$

$\$3,336.13 \times 0.016667 = \$55.60$

$\$3,336.13 + \$55.60 + \$529.74 = \$3,921.47$

Finance charge $= \$55.60$

New balance $= \$3,921.47$

5.8 $12\% \div 365 = 0.0329\% = 0.000329$

$\$692.28 - \$70 = \$622.28$

$\$622.28 \times 0.000329 \times 30 = \$6.14$

$\$622.28 + \$6.14 + \$238.92 = \$867.34$

Finance charge $= \$6.14$

New balance $= \$867.34$

5.9 $4/1, \$1,543.22 \times 2 = \$3,086.44$

$4/3, \$1,543.22 + \$127.89 = \$1,671.11$

$\$1,671.11 \times 2 = \$3,342.22$

$4/5, \$1,671.11 - \$28.13 = \$1,642.98$

$\$1,642.98 \times 3 = \$4,928.94$

$4/8, \$1,642.98 + \$40 = \$1,682.98$

$\$1,682.98 \times 11 = \$18,512.75$

$4/19, \$1,682.98 + \$22.83 = \$1,705.81$

$\$1,705.81 \times 3 = \$5,117.43$

$4/22, \$1,705.81 - \$150 = \$1,555.81$

$\$1,555.81 \times 9 = \$14,002.29$

$\$3,086.44 + \$3,342.22 + \$4,928.94 + \$18,512.78 + \$5,117.43$
$+ \$14,002.29 = \$48,990.10$

$$\frac{\$48,990.10}{30} = \$1,633.00;$$

$18\% \div 12 = 1.5\% = 0.015$

$\$1,633.00 \times 0.015 = \$24.50$

$\$1,543.22 + (\$24.50 + \$127.89 + \$40 + \$22.83)$
$- (\$28.13 + \$150) = \$1,580.31$

Finance charge $= \$24.50$

New balance $= \$1,580.31$

5.10   4/1, $\$1,543.22 \times 4 = \$6,172.88$

4/5, $\$1,543.22 - \$28.13 = \$1,515.09$

$\$1,515.09 \times 17 = \$25,756.53$

4/22, $\$1,515.09 - \$150 = \$1,365.09$

$\$1,365.09 \times 9 = \$12,285.81$

$\$6,172.88 + \$25,756.53 + \$12,285.81 = \$44,215.22$

$$\frac{\$44,215.22}{30} = \$1,473.84;$$

$18\% \div 12 = 1.5\% = 0.015$

$\$1,473.84 \times 0.015 = \$22.11$

$\$1,543.22 + (\$22.11 + \$127.89 + \$40 + \$22.83) - (\$28.13 + \$150) = \$1,577.92$

Finance charge $= \$22.11$

New balance $= \$1,577.92$

5.11 $22\% \div 365 = 0.06030\% = 0.0006030$

$\$300 \times 0.0006030 \times 26 = \$4.70$

$\$10 + \$4.70 = \$14.70$

5.12 $\$400 \times 3\% = \$400 \times 0.03 = \$12$

$21\% \div 365 = 0.0575\% = 0.000575$

$\$400 \times 0.000575 \times 40 = \$9.20$

$\$12 + \$9.20 = \$21.20$

$\$400 + \$21.20 = \$421.20$

5.13

Month	Previous Balance	Finance Charge 1.5% Monthly	New Purchases	Current Balance	Payment $300	Final Balance
January	$2,251.53	$33.77	$0.00	$2,285.30	$300.00	$1,985.30
February	$1,985.30	$29.78	$0.00	$2,015.08	$300.00	$1,715.08
March	$1,715.08	$25.73	$0.00	$1,740.81	$300.00	$1,440.81
April	$1,440.81	$21.61	$0.00	$1,462.42	$300.00	$1,162.42
May	$1,162.42	$17.44	$0.00	$1,179.86	$300.00	$879.86
June	$879.86	$13.20	$0.00	$893.06	$300.00	$593.06
Total					$1,800.00	

Figure A.1
Making $300 payments and no new purchases.

5.14 $\$2,251.13 - \$593.06 = \$1,658.07$

$\$300 \times 6 = \$1,800$

$$\frac{\$1,658.07}{\$1,800} = 0.92 = 92\%$$

Chapter 6

6.1 $I = \$8,000 \times 5\% \times 4 = \$8,000 \times 0.05 \times 4 = \$1,600$

6.2 $I = \$1,000 \times 6.5\% \times 1 = \$1,000 \times 0.065 \times 1 = \$65$

$\$1,000 + \$65 = \$1,065$

6.3 $\quad I = \$4{,}000 \times 3\% \times \dfrac{30}{365} = \$4{,}000 \times 0.03 \times \dfrac{30}{365} = \$9.86$

6.4 $\quad I = \$1{,}500 \times 4.5\% \times \dfrac{60}{360} = \$1{,}500 \times 0.045 \times \dfrac{60}{360} = \$11.25;$

$\$1{,}500 + \$11.25 = \$1{,}511.25$

6.5 $\quad 6\% = 0.06$

$\dfrac{\$2{,}000 \times 0.06}{365} = \$0.3288$

6.6 $\quad \$0.0836 \times 30 = \$2.51$

6.7 $\quad \$30 \times 6 = \$180$

$\$40 + \$180 = \$220$

$\$220 - \$200 = \$20$

$\$20 \div \$200 = 0.1 = 10\%$

Installment price $= \$220$

10% greater than cash price.

6.8 $\quad \$650 - \$50 = \$600$

$\$600 \div 24 = \$25$

6.9 $\quad 12\% \div 12 = 1\% = 0.01$

$\$500 \times 0.01 \times 1 = \$5$

$\$44.42 - \$5 = \$39.42$

$\$500 - \$39.42 = \$460.58$

Interest $= \$5$

Amount applied to principal $= \$39.42$

New balance $= \$460.58$

6.10 $9\% \div 12 = 0.75\% = 0.0075$

$\$4,000 \times 0.0075 \times 1 = \$30$

$\$461.28 - \$30 = \$431.28$

$\$4,000 - \$431.28 = \$3,568.72$

$\$3,568.72 \times 0.0075 \times 1 = \$26.77$

$\$461.28 - \$26.77 = \$434.51$

$\$3,568.72 - \$434.51 = \$3,134.21$

Interest $= \$30, \$26.77$

Amount applied to principal $= \$431.28, \$434.51$

New balance $= \$3,568.72, \$3,134.21$

6.11 $\$415 - \$400 = \$15$

$\$15 \times 12 = \$180;$

$\dfrac{\$180}{\$400} = 0.45 = 45\%$

6.12 $\$50 \times 52 = \$2,600$

$\dfrac{\$2,600}{\$200} = 13 = 1,300\%$

6.13 $\$102.58 \times 72 = \$7,385.76$

$\$7,385.76 - \$5,000 = \$2,385.76$

$\dfrac{2 \times 12 \times \$2,385.76}{(72 + 1) \times \$5,000} = \dfrac{\$57,258.24}{\$365,000} = 0.157 = 15.7\%$

6.14 $\$102.58 \times 72 = \$7,385.76$

$\$7,385.76 + \$85 = \$7,470.76$

$\$7,470.76 - \$5,000 = \$2,470.76$

$\dfrac{2 \times 12 \times \$2,470.76}{(72 + 1) \times \$5,000} = \dfrac{\$59,298.24}{\$365,000} = 0.162 = 16.2\%$

6.15   12% ÷ 12 = 1% = 0.01

$21,117.76 × 0.01 = $211.18

$21,117.76 + $211.18 = $21,328.94

6.16 $556.11 × 60 = $33,366.60

$556.11 × 12 = $6,673.32

$6,673.32 + $21,328.94 = $28,002.26

$33,366.60 − $28,002.26 = $5,364.34

Chapter 7

7.1 $3,000 × 28% = $3,000 × 0.28 = $840

7.2 $5,000 × 28% = $5,000 × 0.28 = $1,400

$5,000 × 36% = $5,000 × 0.36 = $1,800

$1,400 + $600 = $2,000

No, the housing plus other debt is more than 36%.

7.3 $125,000 × 30% = $125,000 × 0.30 = $37,500

$125,000 − $37,500 = $87,500

7.4 $75,000 × 15% = $75,000 × 0.15 = $11,250

$75,000 × 2% = $75,000 × 0.02 = $1,500

$11,250 + $1,500 = $12,750

7.5   30 × 12 = 360

$389.71 × 360 = $140,295.60

7.6 $\$200,000 \times 20\% = \$200,000 \times 0.20 = \$40,000$

$\$200,000 - \$40,000 = \$160,000$

$15 \times 12 = 180$

$\$1,350.17 \times 180 = \$243,030.60$

$\$243,030.60 - \$160,000 = \$83,030.60$

$243,030.60 over the life of the mortgage and pay $83,030.60 in interest.

7.7 $\$875 - \$750 = \$125$

$$\frac{\$2,000}{\$125} = 16$$

7.8 $\$580 - \$500 = \$80$

$$\frac{\$1,500}{\$80} = 18.75;$$

No, she will not break even in 18 months.

7.9 $\$85,000 \div \$100 = 850$

$850 \times \$3.45 = \$2,932.50$

7.10 $\$85,000 \div \$1,000 = 85$

$85 \times \$9.43 = \$801.55$

7.11 $48 \text{ mills} \div 1,000 = \$0.048$

$\$85,000 \times \$0.048 = \$4,080$

7.12 $3.2 \text{ cents} \div 100 = \$0.032$

$\$85,000 \times \$0.032 = \$2,720$

7.13 $\$55,000 \div \$100 = 550$

$550 \times \$0.72 = \$396$

7.14 $6,000 - $500 = $5,500

7.15 $145,000 \times 80\% = $145,000 \times 0.80 = $116,000

$$\frac{$110,000}{$116,000} \times $50,000 = $47,413.79$$

7.16 $6,000 \times 5\% = $6,000 \times 0.05 = $300

$6,000 + $300 + $125 = $6,425

7.17 $23,800 \times 3.5\% = $23,800 \times 0.035 = $833

$23,800 + $833 + $250 + $1,000 - $500 = $25,383

$25,383 - $2,000 = $23,383

7.18 $573.03 \times 48 = $27,505.44

$27,505.44 + $5,000 = $32,505.44

7.19 $32,505.44 - $29,400 = $3,105.44

7.20 $15,000 - $8,500 - $6,500;

$$\frac{$6,500}{4} = $1,625;$$

Total depreciation = $6,500

Average annual depreciation = $1,625

7.21 $$\frac{$1,625}{$15,000} = 0.1083 = 10.83\%$$

7.22 $5 \times 12 = 60;$ $$\frac{$45,000 - $20,000}{60} = \frac{$25,000}{60} = $416.67;$$

($45,000 + $20,000) \times 0.00392 = $65,000 \times 0.00392 = $254.80

$416.67 + $254.80 = $671.47

7.23 $4 \times 12 = 48$; $\dfrac{\$20,000 - \$12,000}{48} = \dfrac{\$8,000}{48} = \$166.67$;

$\dfrac{9.5}{2,400} = 0.00396$

($\$20,000 + \$12,000$) $\times 0.00396 = \$32,000 \times 0.00396 = \$126.72$

$\$166.67 + \$126.72 = \$293.39$

7.24 $100,000 - 85,000 = 15,000$

$15,000 \times \$0.225 = \$3,375$

7.25 $\$239 \times 48 = \$11,472$

$45,000 - 40,000 = 5,000$

$5,000 \times \$0.18 = \$900$

$\$11,472 + \$900 = \$12,372$

7.26 $\$30.88 + \$190.19 = \$221.07$

7.27 $\$324.03 + \$93.99 = \$418.02$

Chapter 8

8.1 $\dfrac{\$500,000}{\$1,000} = 500$;

$500 \times \$1.54 = \$770$

8.2 $\dfrac{\$500,000}{\$1,000} = 500$;

$500 \times \$22.62 = \$11,310$

$\$11,310 - \$770 = \$10,540$

8.3 $\dfrac{\$100,000}{\$1,000} = 100$;

$100 \times 174 = \$17,400$

8.4 $\dfrac{\$1,000,000}{\$1,000} = 1,000;$

$1,000 \times 42 = \$42,000$

8.5 $\$500 \times 60\% = \$500 \times 0.60 = \$300$

$\$500 - \$300 = \$200$

$\$200 \times 12 = \$2,400$

8.6 $\$428 \times 12 = \$5,136$

8.7 $(\$5,000 - \$2,000) \times 15\% = \$3,000 \times 0.15 = \$450$

$\$2,000 + \$450 = \$2,450$

8.8 $\$300 \times 12 = \$3,600$

$(\$9,000 - \$2,000) \times 20\% = \$7,000 \times 0.20 = \$1,400$

$\$3,600 + \$2,000 + \$1,000 + \$1,400 = \$8,000$

8.9 $\$70,000 \times 60\% = \$70,000 \times 0.60 = \$42,000$

$\$42,000 \div 12 = \$3,500$

8.10 $\$3,200 \times 12 = \$38,400$

$\dfrac{\$38,400}{\$70,000} = 0.55 = 55\%$

8.11 $101.275\% = 1.01275$

$\$1,000 \times 1.01275 = \$1,012.75$

8.12 $74.875\% = 0.74875$

$\$1,000 \times 0.74875 = \$748.75$

$\$748.75 \times 10 = \$7,487.50$

8.13 $\$1{,}000 \times 8.4\% \times \dfrac{1}{2} = \$1{,}000 \times 0.084 \times \dfrac{1}{2} = \$42;$

$\$42 \times 10 = \$420$

8.14 $\$1{,}000 \times 7.62\% \times 1 = \$1{,}000 \times 0.0762 \times 1 = \$76.20$

$\$76.20 \times 5 = \$381$

8.15 $\$1{,}000 \times 6\% = \$1{,}000 \times 0.06 = \$60$

$102.458\% = 1.02458$

$\$1{,}000 \times 1.02458 = \$1{,}024.58$

$\dfrac{\$60}{\$1{,}024.58} = 0.059 = 5.9\%$

8.16 $\$53.25 \times 2 = \$106.50$

$93.275\% = 0.93275$

$\$1{,}000 \times 0.93275 = \$932.75$

$\dfrac{\$106.50}{\$932.75} = 0.114 = 11.4\%$

8.17 $112.225\% = 1.12225$

$\$1{,}000 \times 1.12225 = \$1{,}122.25$

$\$1{,}122.25 + \$4 + \$33 = \$1{,}159.25$

$\$1{,}159.25 \times 5 = \$5{,}796.25$

8.18 $88.253\% = 0.88253$

$\$1{,}000 \times 0.88253 = \$882.53$

$\$882.53 + \$125 - \$6 = \$1{,}001.53$

$\$1{,}001.53 \times 10 = \$10{,}015.30$

8.19 $50 \times \$14.80 = \$740$

$\$740 + \$5 = \$745$

8.20 $150 \times \$32.95 = \$4,942.50$

$\$4,942.50 \times 4\% = \$4,942.50 \times 0.04 = \$197.70$

$\$4,942.50 + \$197.70 = \$5,140.20$

8.21 $\$2.38 \times 50 = \$119$

8.22 $\$100 \times 1.8\% = \$100 \times 0.018 = \$1.80$

$\$1.80 \times 200 = \$360$

8.23 $\dfrac{\$336}{\$4,850} = 0.069 = 6.9\%$

8.24 $30 \times \$22 = \$660$

$\$660 \times 2\% = \$660 \times 0.02 = \$13.20$

$\$660 + \$13.20 = \$673.20$

$\$100 \times 3.9\% = \$100 \times 0.039 = \$3.90$

$\$3.90 \times 30 = \$117$

$\dfrac{\$117}{\$673.20} = 0.174 = 17.4\%$

8.25 $200 \times \$9.45 - \$25 = \$1,865$

8.26 $50 \times \$56 = \$2,800$

$\$2,800 \times 1\% = \$2,800 \times 0.01 = \$28$

$\$2,800 + \$28 = \$2,828$

$50 \times \$66 = \$3,300$

$\$3,300 - \$35 = \$3,265$

$\$3,265 + \$60 - \$2,828 = \$497;$

$$\frac{\$497}{\$2,828} = 0.176 = 17.6\%$$

8.27 $\$2,500 \div \$9.82 = 254.582$

8.28 $\$2,500 \div \$10.75 = 232.558$

8.29 $\$6.75 - \$6.03 = \$0.72$

8.30 $\$18.22 - \$17.29 = \$0.93$

8.31 $10,487.33 \times \$10.36 = \$108,648.74$

8.32 $867.508 \times \$6.34 = \$5,500$

$\$5,500 - \$6,000 = -\$500$

$\$500$ loss

8.33 $2.5\% \times 25 = 62.5\%$

$\$86,000 \times 62.5\% = \$86,000 \times 0.625 = \$53,750$

$\$53,750 \div 12 = \$4,479.17$

8.34 $2.5\% \times 28 = 70\%$

$\$92,000 \times 70\% = \$92,000 \times 0.70 = \$64,400$

$\$64,400 \div 12 = \$5,366.67$

$\$5,366.67 - \$4,218.75 = \$1,147.92$ more

8.35 $\$338,000 \div 22.0 = \$15,363.64$

8.36 $\$200,000 \div 17.1 = \$11,695.91$

Chapter 9

9.1 $250 \times 12 = \$3,000$

9.2 $\dfrac{\$28.45 + \$134.27 + 75.63}{3} = \dfrac{\$238.35}{3} = \$79.45$

9.3 $\$1,500 \div 12 = \$125$

$\$955 + \$125 = \$1,080$

9.4 $\$750 \times 12 = \$9,000$

$\dfrac{\$9,000}{\$45,000} = 0.2 = 20\%;$

$\dfrac{\$1,050}{\$3,180} = 0.33 = 33\%;$

$\dfrac{\$500}{\$3,180} = 0.16 = 16\%$

Percent of income saved $= 20\%$

Percent of net pay spent on housing $= 33\%$

Percent of net pay spent on transportation $= 16\%$

Bella spends more on housing than the suggested percentage.

9.5 $\$55,000 \times 20\% = \$55,000 \times 0.20 = \$11,000$

$\$11,000 \div 12 = \$916.67$

9.6 $\$4,000 \times 20\% = \$4,000 \times 0.20 = \$800$

Increase their savings from $600 per month to $800 per month by reducing spending in other categories.

9.7 $\$800 \times 12 = \$9,600$

$\$600 \times 12 = \$7,200$

$\$9,600 - \$7,200 = \$2,400$

9.8 $\dfrac{\$16.50}{125 \text{ oz.}} = \$0.132 \text{ per ounce;}$

$\dfrac{\$9.36}{85 \text{ oz.}} = \$0.110 \text{ per ounce}$

Fresh and Clean is the best buy.

9.9 $36 \div 4 = 9$

$\$5.89 \times 9 = \$53.01$

$\$55.25 - \$53.01 = \$2.24$

$2.24 cheaper to buy 9 packages of 4 batteries.

9.10 $420 \times \$0.08 = \$33.60$

$\$33.60 \times 15\% = \$33.60 \times 0.15 = \$5.04$

$\$33.60 + \$5.04 = \$38.64$

9.11 $420 \div 30 = 14$

$\$10 \times 14 = \$140$

$\$140 \times 15\% = \$140 \times 0.15 = \$21$

$\$140 + \$21 = \$161$

9.12 $\$35 \times 15\% = \$35 \times 0.15 = \$5.25$

$\$35 + \$5.25 = \$40.25$

$\$80 + \$50 = \$130$

$\$130 \div 12 = \$10.83$

$\$40.25 + \$10.83 = \$51.08$

9.13 $\$24.99 \times 6\% = \$24.99 \times 0.06 = \$1.50$

$\$24.99 + \$1.50 + \$10 + \$26 + \$6 = \$68.49$

$\$119.83 - \$68.49 = \$51.34$

9.14 $\$24.99 \times 6\% = \$24.99 \times 0.06 = \$1.50$

$\$24.99 + \$1.50 + \$10 + \$6 = \$42.49$

$\$42.49 + \$14.99 = \$57.48$

Chapter 10

10.1 $\$4,000 \times 7.65\% = \$4,000 \times 0.0765 = \$306$

$\$7,000 - \$4,000 = \$3,000$

$\$3,000 \times 0.8\% = \$3,000 \times 0.008 = \$24$

$\$3,000 \times 5.4\% = \$3,000 \times 0.054 = \$162$

$\$306 + \$24 + \$162 = \$492$

10.2 $\$48,000 \times 7.65\% = \$48,000 \times 0.0765 - \$3,672$

$\$7,000 \times 0.8\% = \$7,000 \times 0.008 = \$56$

$\$7,000 \times 5.4\% = \$7,000 \times 0.054 = \$378$

$\$3,672 + \$56 + \$378 = \$4,106$

10.3 $\$365,000 \times 4\% = \$365,000 \times 0.04 = \$14,600$

$\$365,000 + \$14,600 = \$379,600$

10.4   It will increase the FICA taxes for all employees who are paid less than the Social Security wage limit.

FUTA and SUTA taxes will remain the same for any employee who already made more than \$7,000 per year.

10.5 $70,000 + $10,000 = $80,000

$80,000 ÷ 5,000 = $16

75 × $16 = $1,200

10.6 $250,000 × 10% = $250,000 × 0.10 = $25,000

$25,000 ÷ 12 = $2,083.33

10.7 $250 + $5,500 + $480 = $6,230

$6,230 ÷ 5,000 = $1.25

10.8   60 min = 1 hour

$15.50 ÷ 60 = $0.2583 per minute

8 × $0.2583 = $2.07

10.9 $85,000 + $10,800 = $95,800

$95,800 + $2,500 = $98,300

Prime cost = $95,800

Total manufacturing cost = $98,300

10.10 $98,300 ÷ 25,000 = $3.93

10.11 $\dfrac{\$45,000}{\$25 - \$13} = \dfrac{\$45,000}{\$12} = 3,750$

10.12 $\dfrac{\$45,000}{\$25 - \$10} = \dfrac{\$45,000}{\$15} = 3,000$

10.13 $25,000 \times 15\% = \$25,000 \times 0.15 = \$3,750$

$\$25,000 - \$3,750 = \$21,250$

$\$21,250 \times 15\% = \$21,250 \times 0.15 = \$3,187.50$

$\$21,250 - \$3,187.50 = \$18,062.50$

$\$18,062.50 \times 15\% = \$18,062.50 \times 0.15 = \$2,709.38$

$\$18,062.50 - \$2,709.38 = \$15,353.12$

Book values = $21,250, $18,062.50, $15,353.12

10.14 $25,000 - \$8,000 = \$17,000$

$1 + 2 + 3 + 4 + 5 + 6 + 7 = 28$

$\$17,000 \times \dfrac{7}{28} = \$4,250$

$\$25,000 - \$4,250 = \$20,750$

$\$17,000 \times \dfrac{6}{28} = \$3,642.86$

$\$20,750 - \$3,642.86 = \$17,107.14$

$\$17,000 \times \dfrac{5}{28} = \$3,035.71$

$\$17,107.14 - \$3,035.71 = \$14,071.43$

Book values = $20,750, $17,107.14, $14,071.43

10.15 $25,000 \times 14.29\% = \$25,000 \times 0.1429 = \$3,572.50$

$\$25,000 - \$3,572.50 = \$21,427.50$

$\$25,000 \times 24.49\% = \$25,000 \times 0.2449 = \$6,122.50$

$\$21,427.50 - \$6,122.50 = \$15,305$

$\$25,000 \times 17.49\% = \$25,000 \times 0.1749 = \$4,372.50$

$\$15,305 - \$4,372.50 = \$10,932.50$

Book values = $21,427.50, $15,305, $10,932.50

10.16 3 × $5.38 = $16.14

10.17 $\dfrac{\$500}{\$100}$ × $1.40 = 5 × $1.40 = $7

$12.68 + $7 = $19.68

10.18 $\dfrac{\$4,000}{\$100}$ × $10.75 = 40 × $10.75 = $430

The buyer is responsible.

10.19 Company B because they will pay for the shipping charges.

10.20 700 × $0.52 = $364

$364 × 12 = $4,368

10.21 25,000 × 60% = 25,000 × 0.60 = 15,000

15,000 × $0.50 = $7,500

10.22 $345 + $225 + $65 + $75 = $710

10.23 500 × $0.55 = $275

$30 × 4 = $120

$129 × 3 = $387

$275 + $150 + $387 + $120 = $932

Chapter 11

11.1 $18.25 × 3 = $54.75

$23.99 × 2 = $47.98

$54.75 + $47.98 + $5.62 = $108.35

$108.35 × 5% = $108.35 × 0.05 = $5.42

$108.35 + $5.42 = $113.77

11.2 $63.49 \times 25\% = \$63.49 \times 0.25 = \$15.87$

$63.49 - \$15.87 = \$47.62$

$9.99 + \$47.62 = \$57.61$

$57.61 \times 3.5\% = \$57.61 \times 0.035 = \$2.02$

$57.61 + \$2.02 = \$59.63$

11.3 $250 + \$1,285.36 = \$1,535.36$

$1,535.36 - \$146.92 = \$1,388.44$

$1,388.44 - \$1,387.29 = \$1.15$ short

11.4 $100 + \$2,486.10 = \$2,586.10$

$2,586.10 - \$239.86 = \$2,346.24$

She is even.

11.5 $842.98 + \$622.18 = \$1,465.16$

$1,248.62 + \$1,465.16 - \$2000 = \$713.78$

11.6 $155 \times 20\% = \$155 \times 0.20 = \$31$

$155 - \$31 = \$124$

$124 \times 5.5\% = \$124 \times 0.055 = \$6.82$

$124 + \$6.82 = \$130.82$

$322.18 + \$130.82 = \$453$

11.7 $5,480.62 \times 2\% = \$5,480.62 \times 0.02 = \$109.61$

$5,480.62 - \$109.61 = \$5,371.01$

11.8 The charges are paid within 10 days.

$9,462.89 \times 2\% = \$9,462.89 \times 0.02 = \$189.26$

$9,462.89 - \$189.26 = \$9,273.63$

Shipping charges are paid by seller.

11.9 $\$1{,}050 - \$997.50 = \$52.50$

$$\frac{\$52.50}{\$1{,}050} = 0.05 = 5\%$$

11.10 $\$7{,}500 - \$7{,}237.50 = \$262.50$

$$\frac{\$262.50}{\$7{,}500} = 0.035 = 3.5\%$$

Discount $= \$262.50, \ 3.5\%$

11.11 $\$75 \times 30\% = \$75 \times 0.30 = \$22.50$

11.12 $\$399 \times 35\frac{1}{2}\% = \$399 \times 0.355 = \$141.65$

$\$399 - \$141.65 = \$257.35$

11.13 $\$20 - \$11 = \$9$

$$\frac{\$9}{\$20} = 0.45 = 45\%$$

11.14 $\$9{,}300 \div 150 = \$62$

$\$95.38 - \$62 = \$33.38$

$$\frac{\$33.38}{\$95.38} = 0.35 = 35\%$$

11.15 $\$800 \times 35\% = \$800 \times 0.35 = \$280$

$\$800 - \$280 = \$520$

$\$520 \times 15\% = \$520 \times 0.15 = \$78$

$\$520 - \$78 = \$442$

$\$442 \times 10\% = \$442 \times 0.10 = \$44.20$

$\$442 - \$44.20 = \$397.80$

11.16 $100\% - 45\% = 55\%$

$55\% \times 15\% = 0.55 \times 0.15 = 0.0825 = 8.25\%$

$55\% - 8.25\% = 46.75\%$

$46.75\% \times 5\% = 0.4675 \times 0.05 = 0.023375 = 2.3375\%$

$46.75\% - 2.3375\% = 44.4125\%$

$100\% - 44.125\% = 55.5875\%$

$\$500 \times 55.5875\% = \$500 \times 0.555875 = \$277.94$

Single discount equivalent $= 55.875\%$

Discount $= \$277.94$

11.17 $\$225 \times 55\% = \$225 \times 0.55 = \$123.75$

$\$225 + \$123.75 = \$348.75$

Markup $= \$123.75$

Selling price $= \$348.75$

11.18 $\$25.56 - \$18 = \$7.56$

$$\frac{\$7.56}{\$18} = 0.42 = 42\%$$

11.19 $\dfrac{\$8}{0.45} = \dfrac{\$8}{45\%} = \$17.78$

$\$17.78 + \$8 = \$25.78$

Cost $= \$17.78$

Selling price $= \$25.78$

11.20 $\$99.99 \times 40\% = \$99.99 \times 0.40 = \$40.00$

$\$99.99 - \$40 = \$59.99$

11.21 $96 - $48 = $48

$$\frac{\$48}{\$96} = 0.50 = 50\%$$

11.22 $\dfrac{\$50}{40\%} = \dfrac{\$50}{0.40} = \$125$

$125 - 50 = $75

Cost = $75

Selling price = $125

11.23 $39.98 × 30% = $39.98 × 0.30 = $11.99

$39.98 - $11.99 = $27.99

11.24 $250 × 35% = $250 × 0.35 = $87.50

$250 - $87.50 = $162.50

$162.50 × 45% = $162.50 × 0.45 = $73.13

$162.50 + $73.13 = $235.63

$235.63 × 15% = $235.63 × 0.15 = $35.34

$235.63 - $35.34 = $200.29

$200.29 × 40% = $200.29 × 0.40 = $80.12

$200.29 - $80.12 = $120.17

Final sales price = $120.17, which does not cover the $162.50 the retailer paid for the tent.

Chapter 12

12.1 $130,000 \times 35\% = \$130,000 \times 0.35 = \$45,500$

$275,000 \times 15\% = \$275,000 \times 0.15 = \$41,250$

$125,000 \times 25\% = \$125,000 \times 0.25 = \$31,250$

$45,500 + \$41,250 + \$31,250 = \$118,000$

12.2 $118,000 \div 5,000 = \$23.60$

12.3 $165 - 80 + 150 - 50 - 85 + 200 - 125 = 175$

12.4 $1,800 + 250 + 1,000 + 1,000 - 150 - 580 - 1,250 - 487 = 1,583$

12.5 $\dfrac{150 \text{ gallons}}{30 \text{ days}} = 5 \text{ gallons per day}$

$(5 \times 4) + 15 = 20 + 15 = 35$

12.6 $\dfrac{5,000 \text{ belts}}{25 \text{ days}} = 200 \text{ belts per day}$

$200 \times 3 = 600$

$(200 \times 6) + 600 = 1,200 + 600 = 1,800$

12.7 $50 \times \$11 = \$550$

$75 \times \$10.50 = \$787.50$

$50 + 75 = 125$

$175 - 125 = 50$

$50 \times \$9 = \$450$

$550 + \$787.50 + \$450 = \$1,787.50$

12.8 $175 \times \$7 = \$1,225$

12.9 $450 \times \$7 = \$3,150$

$125 \times \$7.50 = \$937.50$

$200 \times \$8.25 = \$1,650$

$250 \times \$9 = \$2,250$

$75 \times \$10.50 = \$787.50$

$50 \times \$11 = \$550$

$\$3,150 + \$937.50 + \$1,650 + \$2,250 + \$787.50 + \$550 = \$9,325$

$450 + 125 + 200 + 250 + 75 + 50 = 1,150$

$\$9,325 \div 1,150 = \$8.11$

$175 \times \$8.11 = \$1,419.25$

12.10 $\$550,000 \times 9\% = \$550,000 \times 0.09 = \$49,500$

$\$550,000 \times 5\% = \$550,000 \times 0.05 = \$27,500$

$\$550,000 \times 2\% = \$550,000 \times 0.02 = \$11,000$

$\$550,000 \times 8\% = \$550,000 \times 0.08 = \$44,000$

$\$49,500 + \$27,500 + \$11,000 + \$44,000 + \$45,000 = \$177,000$

12.11 $\$550,000 \times 25\% = \$550,000 \times 0.25 = \$137,500$

Chapter 13

13.1 $\$200,000 - \$2,675 = \$197,325$

13.2 $\$527,836 + \$302,400 - \$638,525 = \$191,711$

13.3 $\$1,847,500 + \$2,834,765 - \$1,655,150 = \$3,027,115$

13.4 $1,487,950 - $338,460 = $1,149,490$

13.5 $985,462 - $18,500 = $966,962$

$653,828 + $250,595 - $515,787 = $388,636$

$966,962 - $388,636 = $578,326$

13.6 $12,500 + $1,500 + $720 + $580 + $225 + $310 + $142 + $512 = $16,489$

$42,876 - $16,489 = $26,387$ net income

13.7

Revenue		
Sales	215,000	
Returns and allowances	9,026	
Net sales		205,974
Cost of Goods Sold		
Beginning inventory	23,179	
Cost of goods manufactured	158,107	
Goods available for sale	181,286	
Ending inventory	20,486	
Cost of goods sold		160,800
Gross profit		45,174
Operating expenses		
Salaries	19,487	
Office expenses	3,256	
Payroll taxes	2,612	
Depreciation	2,788	
Total operating expenses		28,143
Net Income		17,031

13.8   Assets = \$15,842.92 + \$62,946.33 + \$539,429.22 + \$6,482
+ \$42,156 + \$175,000 = \$841,856.47

Liabilities = \$285,936 + \$85,000 = \$370,936

Equity = \$841,856.47 − \$370,936 = \$470,920.47

13.9   Assets = \$7,260 + \$15,492 + \$35,844 + \$1,225 = \$59,821

Liabilities = \$14,497 + \$9,500 = \$23,997

Equity = \$59,821 − \$23,997 = \$35,824

13.10 $\dfrac{\$45,174}{\$205,974} = 0.219 = 21.9\%$

13.11 $\dfrac{\$17,031}{\$205,974} = 0.083 = 8.3\%$

13.12 $\dfrac{\$1,863,225 + \$1,563,480}{2} = \dfrac{\$3,426,705}{2} = \$1,713,352.50;$

$\dfrac{\$30,840,350}{\$1,713,352.50} = 18.0$

13.13 $\dfrac{\$68,000 + \$72,585}{2} = \dfrac{\$140,585}{2} = \$70,292.50;$

$\dfrac{\$630,000}{\$70,292.50} = 9.0$

13.14   \$15,842.92 + \$62,946.33 + \$539,429.22 + \$6,482 = \$624,700.47

$\dfrac{\$624,700}{\$285,936} = 2.2;\ 2.1{:}1$

13.15 $7,260 + \$15,492 + \$1,225 = \$23,977$

$$\frac{\$23,977}{\$14,497} = 1.7; \ 1.7{:}1$$

13.16 Assets $= \$15,842.92 + \$62,946.33 + \$539,429.22 + \$6,482 +$
$\$42,156 + \$175,000 = \$841,856.47$

Liabilities $= \$285,936 + \$85,000 = \$370,936$

Equity $= \$841,856.47 - \$370,936 = \$470,920.47$

$$\frac{\$370,936}{\$470,920.47} = 0.788 = 78.8\%$$

13.17 Assets $= \$7,260 + \$15,492 + \$35,844 + \$1,225 = \$59,821$

Liabilities $= \$14,497 + \$9,500 = \$23,997$

Equity $= \$59,821 - \$23,997 = \$35,824$

$$\frac{\$23,997}{\$35,824} = 0.670 = 67.0\%$$

13.18 $$\frac{\$125,600}{\$571,000} = 0.220 = 22.0\%$$

13.19 $482,955 - \$257,800 = \$225,155$

$225,155 - \$122,729 = \$102,426$

$$\frac{\$102,426}{\$355,823} = 0.288 = 28.8\%$$

Index